Disposable and Flexible Chemical Sensors and Biosensors Made with Renewable Materials

Disposable and Flexible Chemical Sensors and Biosensors Made with Renewable Materials

Editor

Jaehwan Kim

Inha University, South Korea

World Scientific

NEW JERSEY · LONDON · SINGAPORE · BEIJING · SHANGHAI · HONG KONG · TAIPEI · CHENNAI · TOKYO

Published by

World Scientific Publishing Europe Ltd.

57 Shelton Street, Covent Garden, London WC2H 9HE

Head office: 5 Toh Tuck Link, Singapore 596224

USA office: 27 Warren Street, Suite 401-402, Hackensack, NJ 07601

Library of Congress Cataloging-in-Publication Data

Name: Kim, Jaehwan, 1961– editor.

Title: Disposable and flexible chemical sensors and biosensors made with renewable materials / edited by Jaehwan Kim (Inha University, South Korea).

Description: New Jersey : World Scientific, [2017] | Includes bibliographical references.

Identifiers: LCCN 2017011378 | ISBN 9781786343864 (hc : alk. paper)

Subjects: LCSH: Chemical detectors--Materials. | Biosensors--Materials. | Flexible electronics. | Renewable natural resources.

Classification: LCC TP159.C46 K56 2017 | DDC 681/.2--dc23

LC record available at https://lccn.loc.gov/2017011378

British Library Cataloguing-in-Publication Data

A catalogue record for this book is available from the British Library.

Desk Editors: Anthony Alexander/Jennifer Brough/Koe Shi Ying

Typeset by Stallion Press
Email: enquiries@stallionpress.com

Preface

Most sensors are based on ceramic or semiconducting substrates that have no flexibility or biocompatibility. Polymer-based sensors are the subject of much attention due to their ability to collect molecules on their sensing surface with flexibility. However, most petroleum-based polymers are nonrenewable, thereby causing problems in disposal of sensors. Beyond polymer substrates for sensors, renewable materials-based substrates for sensors will bring technological impacts in many areas due to their environmentally friendly, flexible, and disposable characteristics. This book reviews current state-of-the-art of renewable material-based chemical sensors and biosensors, and their potential applications are suggested for industrial, environmental, and biomedical areas.

So far, many books have been published for chemical sensors and biosensors, and renewable material sensors have been published separately from chemical sensors and biosensors. This book bridges two interdisciplinary areas and envisages its potential for the future. In Chapter 1, the reason why flexible sensors and disposable sensors are necessary is explained, and their applications in terms of biomedical care, smart packaging, and smart city are illustrated. Chapter 2 introduces renewable biopolymers, classified into two groups: natural polymers directly derived from biosources, e.g., polysaccharides, protein, lipid, and poly(hydroxyalkanoates), and chemically synthetic polymers utilizing natural monomers, e.g., poly(lactic acid) and biopolyurethane. This chapter discusses these biopolymers and summarizes their sources, chemical structures, properties, and potential applications in diverse fields. Chapter 3 explains all about sensor principles including resistive-type sensors, capacitive-type sensors, and impedance-type sensors. Chapter 4 introduces chemical sensors and includes recent developments on receptor materials, paper-based sensors, biofriendly disposable sensors, pH sensors, colorimetric sensors, and electronic nose. Chapter 5 explains the basics of biosensors, including bioreceptors and transducers, and outlooks of biosensors for paper-based disposable sensors are introduced.

The editor of this book would like to express gratitude to all contributing authors for their valuable chapters that deliver fundamentals of and insights into flexible and disposable sensors to readers such as professors, graduate students, and researchers in the areas of biotechnology, nanotechnology, bioengineering, biomedical engineering, material science, mechanical engineering, polymer engineering, and medical schools. Finally, the editor would like to thank the editorial staff of this book at Imperial College Press for their great effort and support for publication.

All the best,

Jaehwan Kim, Ph.D.
Director, Creative Research Center for Nanocellulose
Future Composites
Inha Fellow Professor, Department of Mechanical Engineering
Inha University
Incheon 22212, Republic of Korea
Jaehwan@inha.ac.kr

About the Editor

Jaehwan Kim received his B.S. in Mechanical Engineering from Inha University in 1985, M.S. in Mechanical Engineering from KAIST in 1987, and Ph.D. in Engineering Science and Mechanics from The Pennsylvania State University in 1995. In March 1996, he joined the Department of Mechanical Engineering at Inha University, Korea, where he serves as an Inha Fellow Professor.

Dr. Kim is a Fellow of The Korean Academy of Science and Technology, National Academy of Engineering of Korea, and Institute of Physics. He is an Associate Editor of *Smart Materials and Structure* as well as *Smart Nanosystems in Engineering and Medicine* and Editor of *International Journal of Precision Manufacturing and Engineering, Helyon and Actuators*. He has been the Director of Creative Research Center for EAPap Actuator funded by National Research Foundation of Korea (NRF). Recently, he started the Creative Research Center for Nanocellulose Future Composites, sponsored by NRF.

He first discovered cellulose as a smart material which can be used in sensors, actuators and electronic materials. His research interests are smart materials, structures and devices, biomaterial-based smart materials, cellulose, electroactive polymers, power harvesting, biomimetic actuators, biosensors, tactile sensors, and flexible electronics. He has published more than 260 prestigious journal papers, presented 290 international conference papers, and filed more than 30 patents.

About the Authors

Hyun Chan Kim received his B.S. from the Department of Mechanical Engineering, Inha University, South Korea, in 2014. He is now a Ph.D. student in the Department of Mechanical Engineering at Inha University, South Korea. His research interest is the fabrication of nanocellulose-based composites and its characterization.

Jung Woong Kim received his B.S. from the Department of Mechanical Engineering, Inha University, South Korea, in 2014. He is now a Ph.D. student in the Department of Mechanical Engineering at Inha University, South Korea. His research interest lies in nanocellulose-based composites and its space applications.

Young-Jun Lee received his B.S. and M.E. from the Department of Electronic Engineering, Chosun University, Gwangju, South Korea, in 2011 and 2013, respectively. He is currently pursuing his Ph.D. in the Department of Mechanical Engineering, Inha University, South Korea. His research interests are thin film devices and microsensor systems with functional materials.

Gwang-Wook Hong received his B.S. from the Department of Electrical Engineering, Chosun University, South Korea in 2013, and M.E. from the Department of Mechanical Engineering, Inha University, South Korea, in 2015. He is currently a Ph.D. student in the Department of Mechanical Engineering, Inha University. His research interests are advanced 3D printing techniques, infrared thermography, and flexible sensing device for human monitoring.

Joo-Hyung Kim received his B.S. and M.E. from the Department of Mechanical Engineering, Inha University, South Korea, in 1993 and 1995, respectively, and his Ph.D. from the Department of Microelectronics and Information Technology, KTH Royal Institute of Technology, Stockholm, Sweden, in 2005. From 1995 to 2002, he was a Senior Research Engineer with Daewoo and

Samsung SDI central research centers, Korea. From 2006 to 2008, he was a Senior Scientist with the Fraunhofer Institute, Germany, for novel material research in microelectronics. He was an Assistant Professor in Department of Electronic Engineering, Chosun University. Currently he is an Associate Professor in the Department of Mechanical Engineering, Inha University, South Korea. His research interests are microelectromechanical systems, sensors, and advanced 3D printing and advanced smart mechanical systems.

Kyungbae Woo received his B.S. from the Department of Chemical Engineering and Biological Engineering in Inha University, South Korea, in 2016. His research area is natural conductivity material composites for biomedical applications.

Sangkyu Lee received his B.S. (2012) and M.E. (2014) from the Department of Chemical Engineering in Inha University, South Korea. He is currently a material research team researcher at Daehan Solution.

Jae Eun Heo received her B.S. from the Department of Chemical Engineering in Inha University, South Korea, in 2015. Her research is focused on the various types of cellulose composites.

Seunghyeon Lee is currently pursuing B.S. in the Department of Chemical Engineering in Inha University, South Korea. His minor specialization is biological sciences and his research area is natural composites for biomedical applications.

Daseul Jang received her B.S. and M.E. from the Department of Chemical Engineering in Inha University, South Korea. Her research interests focus on synthesis of bioinspired nanocomposites, high mechanical performance composites, and bio-based polymers.

Bong Sup Shim is currently an Associate Professor of Chemical Engineering at Inha University, Korea. He has received his Ph.D. in Chemical Engineering at the University of Michigan, followed by postdoc training at the University of Delaware. His research interest includes nanocomposites, biocomposites, biomimetic materials, neural interfaces, drug delivery and artificial organs.

Contents

Chapter 1

Introduction

Jaehwan Kim*, Jung Woong Kim and Hyun Chan Kim

Creative Research Center for Nanocellulose Future Composites
Department of Mechanical Engineering, Inha University
Inha-Ro 100, Nam-Ku, Incheon 22212, Republic of Korea
*jaehwan@inha.ac.kr

1. Why Flexible and Disposable Sensors?

The importance of sensors is increasing as our society falls further under the influence of the internet of things (IoT). IoT supports technology frames for home automation, smart cities, medical care, smart packaging, and wearable electronics by utilizing ubiquitous sensors. The IoT sensor market is expected to reach USD 38.41 billion by 2022, growing at a compound annual growth rate (CAGR) of 42.08% between 2016 and 2022.[1] The need for IoT sensors has driven the development of cheaper, smarter, and smaller sensors for smart and wearable devices. Through the advancement of nanotechnology, biotechnology, information technology, and semiconductor technology, these challenging developments have been possible. Over and above these challenges, however, since our society faces problems of pollution, shortage of natural resources including oil and energy, and aging, the technologies under development should be harmonized with our society and nature. So far, most sensors are based on ceramic or semiconducting substrates, which have no flexibility or biocompatibility. Polymer-based sensors are the subject of much attention due to their ability to flexibly collect molecules on their sensory surfaces. However, most petroleum-based polymers are not renewable, thereby causing problems in their disposal. To be harmonized with nature, materials for sensors need to be renewable materials, for example, paper.[2,3] By using renewable materials, sensors can be made flexible, disposable, and cheap. Renewable material-based substrates for sensors will bring technological impacts in many areas due to their environmentally friendly, flexible, and disposable characteristics. Various sensors can be fabricated on paper or renewable material substrates by means of efficient state-of-the-art production technologies, for example, roll-to-roll printing, inkjet printing, and 3D printing. This will be an important trend of future technologies with broad technological impacts in many areas such as medical care, packaging, smart cities, and so on. Figure 1.1 shows the concept of flexible and disposable sensors for the sustainability of our society. Sensors for medical care and smart packaging are briefly introduced in the following sections.

Fig. 1.1.   Concept of flexible and disposable sensors for human–nature sustainability.

2. Sensors for Medical Care

Sensors are used in electronics-based medical equipment to convert various forms of stimuli into electrical signals for analysis. Sensors can increase the intelligence of medical equipment, such as life-supporting implants, and can enable bedside and remote monitoring of vital signs and other health factors. With a growing aged population, design of responsive, noninvasive, and user-friendly medical devices, which can cater to the needs of an older generation, is essential. Patients are looking for cost-effective and easy-to-use assistive tools that help them regain independence and confidence in their everyday life. The goal of medical assistive devices is for patients to regain or maintain their independence and grow confidence in their ability to function to their own expectations or those of a physician. An aging and expanding population is accelerating the development of new and different types of medical equipment, including various sensors used inside both equipment and patients' bodies. Hospitals want real-time, reliable, and accurate diagnostic results provided by the devices that can monitor remotely, whether the patient is in a hospital, clinic, or at home. These devices must be affordable and user-friendly. Today, most devices are technologically savvy, implementing some kind of wireless and/or Bluetooth feedback system. Feedback is the key. The device needs to be responsive and provide the user or doctor with unique data. This data is used to eliminate guesswork and allows the doctor to make diagnoses that are more justified. Advances in radio-frequency identification (RFID) and sensor technologies' ability to communicate with each other are influencing research and development in many areas, such as supply chain, the military, homeland security, civil infrastructure, traffic control, and medical applications. As sensors and RFIDs become increasingly interoperable, manufacturers are better able to monitor the location and condition of objects. Real-time monitoring of vital signs and movement, on-demand drug delivery systems, and continuous blood analysis are very important to maintain for patients' safety.[4] Growing needs in the point-of-care, which is an interesting market for improving patients' quality of life, drive the development of sensor technologies for diagnosis and treatment of various life-threatening diseases. Current remote sensor technology will allow real-time measurement of body temperature, respiratory rate, heart rate, and electrocardiogram (ECG) signal, and eventually blood pressure and blood gas concentrations.

A bed monitoring system is very useful for patients, especially geriatrics. Monitoring systems are a way for doctors, patients, and their loved ones to keep constant track of a patient's day-to-day activities in the comfort of their own natural environment.[6] New medical devices allow doctors 24-hour access to this insightful information regarding a patient's progression and everyday routine. Real-time monitoring systems are becoming increasingly popular in geriatrics, specifically keeping a close eye on elderly patients. Force sensors can be integrated into wearable medical devices as well as stationary furniture so as to react to the need of patients and make them comfortable. This is a form of smart medical electronics, the so-called Smart Meditronics. Smart Meditronics can sense a patient's condition and environment changes in such a way that it can immediately report the information to the doctor, in real time. Appropriate responses are programed in order to make patients comfortable. To be able to implement the functions of Smart Meditronics, sensors and actuators in conjunction with control electronics should be integrated. Figure 1.2 represents the concept of Smart Meditronics.

3. Sensors for Smart Packaging

These sensing and responding networks can help improve the security, quality, and integrity of products moving through the supply chain, including perishable foods, temperature-sensitive high-tech equipment, hazardous materials, and equipment prone to humidity-induced erosion.[6] This is the so-called smart packaging. Smart packaging is an integrated technology of delivery monitoring and packaging along with information technology. Smart packaging can inform safety and delivery status of products to customers conveniently, for example, freshness of agricultural products and safety of organs or medical products. Smart packaging provides enhanced

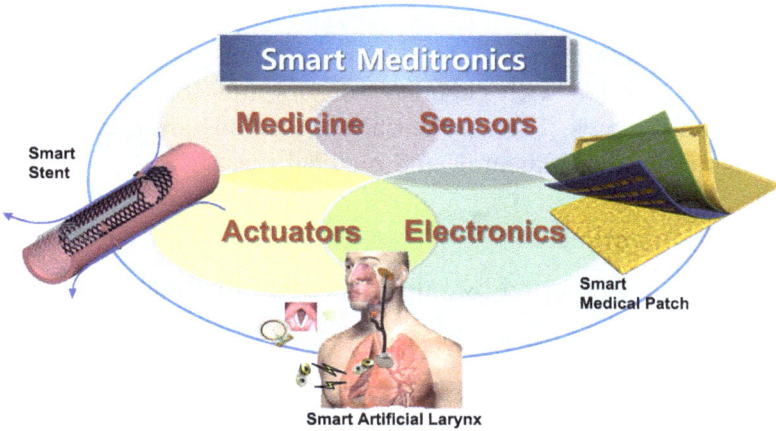

Fig. 1.2.   Concept of Smart Meditronics.

functionality that can be divided into two submarkets: active packaging, which provides functionality such as moisture control, and intelligent packaging, which incorporates features that indicate status or communicate product changes and other information.[7] To meet the functions for smart packaging and various sensors such as temperature sensors, humidity sensors, biosensors, chemical sensors, and pH sensors, RFID tags along with paper batteries and flexible electronics should be fabricated on a flexible paper.[8–28] Since this kind of flexible and disposable sensors are made on paper substrates, it will be beneficial for achieving sustainable technology.

4. Summary

Most sensors are based on ceramic or semiconducting substrates, which have no flexibility or biocompatibility. Polymer-based sensors gain much attention due to their ability to collect molecules on their sensing surface with flexibility. However, most petroleum-based polymers are not renewable, thereby causing a problem in the disposal of sensors. Beyond polymer substrates for sensors, renewable materials-based substrates for sensors will bring technological impacts in many areas, for example, medical care, smart packaging, and smart city due to their environmentally friendly, flexible, and disposable character-istics. This book reviews current state-of-the-art of renewable materials in conjunction with chemical sensors and biosensors made of renewable materials. Their potential applications are also illustrated for industrial, environmental, and medical areas.

Acknowledgement

This work was supported by Inha University Research Grant.

References

1. http://www.marketsandmarkets.com/Market-Reports/sensors-iot-market-26520972.html.
2. J.-H. Kim, B. S. Shim, H. S. Kim, Y.-J. Lee, S.-K. Min, D. Jang, Z. Abas and J. Kim, Review of nanocellulose for sustainable future materials, *J. Precision Eng. Manufact. — Green Technol.* **2**(2), 197–213 (2015).
3. H. C. Kim, S. Mun, H.-U. Ko, L. Zhai, A. Kafy and J. Kim, Renewable smart materials, *Smart Mater. Struct.* **25**, 073001 (14pp) (2016).
4. K. Townsend, J. Haslett, T. K. K. Tsang, M. N. El-Gamal and K. Iniewski, Recent advances and future trends in low power wireless systems for medical applications, In *Proc. 5th IEEE International Workshop on System-on-Chip for Real-Time Applications*, pp. 20–24, Banff, Canada (July, 2005).
5. Tekscan, Assisting an aging population: Designing medical devices with force sensing technology, https://www.tekscan.com/resources/whitepaper/assisting-aging-population-designing-medical-devices-force-sensing-technology.
6. C. Stephenson, Sense and respond networks for agile, secure distribution, *Sens. Magaz.* **23**(9), 14–18 (2006).
7. http://www.packagingdigest.com/smart-packaging.
8. K. K. Sadasivuni, A. Kafy, H.-C. Kim, H.-U. Ko, S. Mun and J. Kim, Reduced graphene oxide filled cellulose films for flexible temperature sensor application, *Synth. Met.* **206**, 154–161 (2015).
9. A. Kafy, A. Akther, M. I. R. Shishir, H. C. Kim, Y. Yun and J. Kim, Cellulose nanocrystal/graphene oxide composite film as humidity sensor, *Sens. Actuat. A.* **247**, 221–226 (2016).
10. S. K. Mahadeva, S. Yun and J. Kim, Flexible humidity and temperature sensor based on cellulose-polypyrrole nanocomposite, *Sens. Actuat. A.* **165**(2), 194–199 (2011).
11. K. K. Sadasivuni, D. Ponnamma, H.-U. Ko, H. C. Kim, L. Zhai and J. Kim, Flexible NO_2 sensors from renewable cellulose nanocrystals/iron oxide composites, *Sens. Actuat. B.* **233**, 633–638 (2016).

12. A. Kafy, K. K. Sadasivuni, A. Akther, S.-K. Min and J. Kim, Cellulose/ graphene nanocomposite as multifunctional electronic and solvent sensor material, *Mater. Lett.* **159**, 20–23 (2015).

13. J.-H. Kim, S. Mun, H.-U. Ko, G.-Y. Yun and J. Kim, Disposable chemical sensors and biosensors made on cellulose paper, *Nanotechnol.* **25**(9), 092001(7pp) (2014).

14. Y. Chen, S. Mun and J. Kim, A wide range conductometric pH sensor made with titanium dioxide/multiwall carbon nanotube/cellulose hybrid nanocomposite, *IEEE Sens. J.* **13**(11), 4157–4162 (2013).

15. S. K. Mahadeva, H.-U. Ko and J. Kim, Investigation of cellulose and tin oxide hybrid composite as a disposable pH sensor, *Z. Phys. Chem.* **227**(4) 419–428 (2013).

16. S. Mun, Y. Chen and J. Kim, Cellulose-titanium dioxide-multiwalled carbon nanotube hybrid nanocomposite and its ammonia gas sensing properties at room temperature, *Sens. Actuat. B.* **171–172**, 1186–1191 (2012).

17. Y. Chen, S.-D. Jang and J. Kim, Gas sensing properties of gallium nitride-coated cellulose nanocomposite, *Sensor Lett.* **10**(3–4), 748–753 (2012).

18. S. Yun and J. Kim, Multi-walled carbon nanotubes-cellulose paper for a chemical vapor sensor, *Sens. Actuat. B.* **150**, 308–313 (2010).

19. A. Kafy, M. Maniruzzaman, B.-W. Kang and J. Kim, Fabrication and Characterization of titanium dioxide-cellulose composite and its urea biosensing behavior, *Sens. Mater.* **27**(7), 539–548 (2015).

20. S. Mun, M. Maniruzzaman, H.-U. Ko, A. Kafy and J. Kim, Preparation and characterization of cellulose ZnO hybrid film by blending method and its glucose biosensor application, *Mater. Technol.: Adv. Biomater.* **2**(3), B150–B154 (2015).

21. S. K. Mahadeva, B.-W. Kang and J. Kim, Detection of urea and rancidity of milk using inter-digitated cellulose-tin oxide hybrid composite, *Sens. Lett.* **12**(1), 39–43 (2014).

22. S. K. Mahadeva and J. Kim, Porous tin-oxide-coated regenerated cellulose as disposable and low-cost alternative transducer for urea detection, *IEEE Sens. J.* **13**(6), 2223–2228 (2013).

23. M. Maniruzzaman, S.-D. Jang and J. Kim, Titanium dioxide-cellulose hybrid nanocomposite and its glucose biosensor application, *Mater. Sci. & Eng. B.* **177**(11), 844–848 (2012).

24. S. K. Mahadeva and J. Kim, Conductometric glucose biosensor made with cellulose and tin oxide hybrid nanocomposite, *Sens. Actuat. B.* **157**, 177–182 (2011).

25. A. Kafy, K. K. Sadasivuni, H.-C. Kim, A. Akther and J. Kim, Designing flexible energy and memory storage materials using cellulose modified graphene oxide nanocomposites, *Phys. Chem. — Chem. Phys.* **17**, 5923–5931 (2015).

26. J.-H. Kim, S. Yun, H.-U. Ko and J. Kim, A flexible paper transistor made with aligned single-walled carbon nanotube bonded cellulose composite, *Curr. Appl. Phys.* **13**(5), 897–901 (2013).

27. S. Yun, S.-D. Jang, G.-Y. Yun, J.-H. Kim and J. Kim, Paper transistor made with covalently bonded multiwalled carbon nanotube and cellulose, *Appl. Phys. Lett.* **95**(10), 104102(3pp) (2009).
28. J. Kim, S. Yun and Z. Ounaies, Discovery of cellulose as a smart material, *Macromol.* **39**, 4202–4206 (2006).

Chapter 2

Renewable Materials

Daseul Jang, Kyungbae Woo and Bong Sup Shim*

Department of Chemical Engineering, Inha University
100 Inha-Ro, Incheon 22212, Republic of Korea
*bshim@inha.ac.kr

Demands for replacing petro-materials have grown due to environmental concerns, depletion of petro-sources, and market potential for environmentally friendly materials. Biopolymers, which inherently possess the properties of biodegradability, low toxicity, and carbon neutrality, are attractive alternatives for the conventional polymers because they are easily extracted from abundant and inexhaustible natural sources. Renewable biopolymers are classified into two groups. One is natural polymers directly derived from biosources such as plants, animals, and microorganisms (e.g., polysaccharides, protein, lipid, and poly(hydroxyalkanoates) (PHAs)). The others are chemically synthetic polymers utilizing natural monomers such as amino acids, sugars, fats, and oils (e.g., poly(lactic acid) (PLA), and bio-polyurethane (bio-PU)). This chapter discusses these biopolymers and summarizes their sources, chemical structures, properties, and potential applications in diverse fields such as materials science, biomedical engineering, and electronics.

1. Introduction: Materials from Renewable Sources

Materials from renewable sources receive increasing attention as fossil fuel resources cause environmental degradation not just from their exhaustion but also from atmospheric changes such as CO_2 accumulation and pollution. In our modern daily life, numerous commodities and goods are made with petroleum-based chemicals and materials. However, petroleum-based chemicals involve environmentally open circuit cycles. The more petroleum-materials are mined and used, ultimately the more carbon dioxide is accumulated on earth. This irreversible process will cause significantly destructive problems to our existence. Furthermore, petroleum-based materials are not biodegradable, so waste accumulates in the soil and oceans. Although the price of crude oil was temporarily low in 2016, unpredictable price variations adds pressure to replacing petroleum-based materials. Therefore, pollution, environmental changes, depletion of petroleum sources, and fluctuations of oil prices, all draw attention to the need for converting petroleum-based chemicals and materials to renewable and sustainable ones which are capable of biodegradability and carbon neutrality (Fig. 2.1).[1]

The term biopolymers is used to refer to several classes of polymers, such as biodegradable, biocompatible, or bioorigin polymers. Among them, only bioorigin polymers can be considered renewable materials. Renewable polymers are categorized into natural polymers, which are produced and stored directly from biological systems

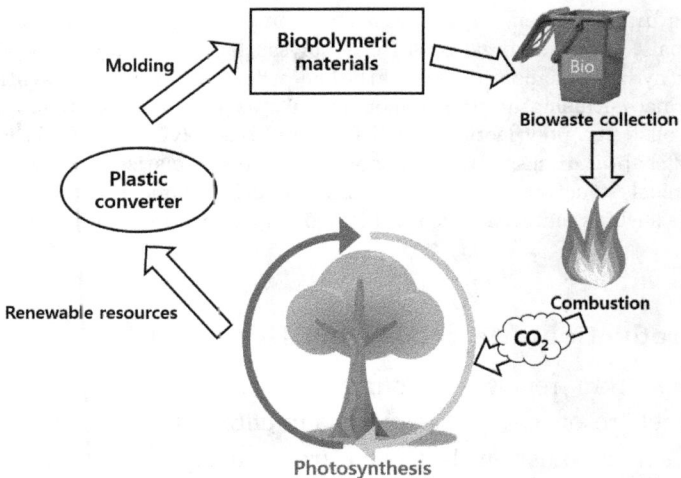

Fig. 2.1. Cycle of carbon in biopolymer-based materials. (Redesigned the following concept of Ref. 1).

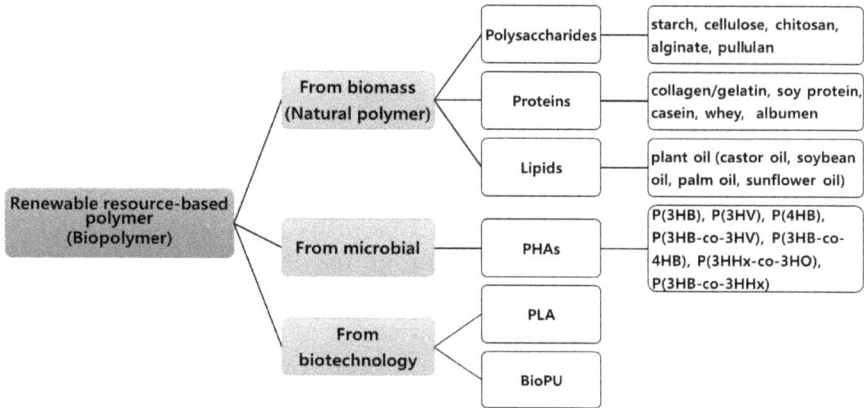

Fig. 2.2.   Classification of biopolymers.

of animals, plants, and microorganisms, and chemically synthesized polymers from natural chemicals such as amino acids, sugars, natural fats, and oils, as shown in Fig. 2.2.[2,3] In the following sections, we discuss these biopolymers in detail.

2. Polysaccharides

Polysaccharides are naturally occurring polymers including starch, cellulose, chitin, and alginate, which consist of monosaccharides (sugars) linked by glycosidic bonds. Monosaccharides have cyclic backbone with five or six carbons, such as ribose, glucose, and galactose. Repeating units of polysaccharides are composed of more than one monosaccharide. Their structures can be diverse according to their molecular arrangement. For instance, dextran is a branched polymer with a number of glucose repeating units.[2] Polysaccharides are unique raw materials with the following advantageous features[4,5]:

- They are abundant, obtainable everywhere on earth, and inexpensive.
- They have intrinsically high stability, non-toxicity, biodegradability, biocompatibility, hydrophilic surfaces, and high mechanical properties, particularly because of robust hydrogen bonds.

- They are inherently hydrophilic and hygroscopic. However, their solubility in a number of solvents can be easily controlled by modifying molecular functional groups.

2.1. *Starch*

Starch is an abundant, widely available, and cheap semicrystalline biopolymer extracted from grains such as potato, maize (corn), wheat, and rice.[6] This material is biosynthesized in the form of granules (Fig. 2.3), consisting of two different macromolecules, linear amylose (poly-α-1,4-D-glucopyranoside) and highly branched amylopectin (poly-α-1,4-D-glucopyranoside and poly-α-1,6-D-glucopyranoside) (Fig. 2.4).[7,8] The ratio of amylose and amylopectin varies based on the origin of the starch (Table 2.1). This

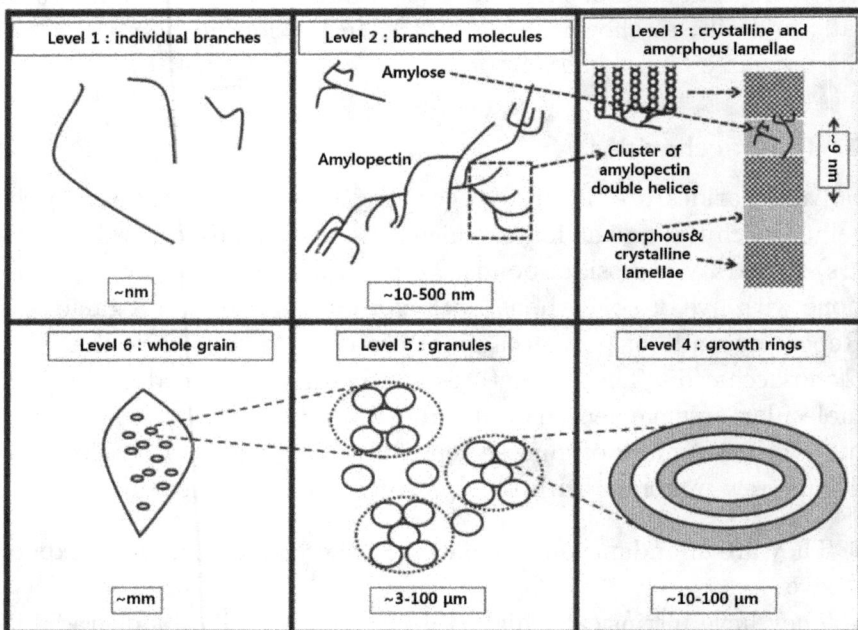

Fig. 2.3.   Structure of starch from macro- to nano-level.[7] (Reproduced from Wu *et al.* (2013) with permission from CSIRO Publishing).

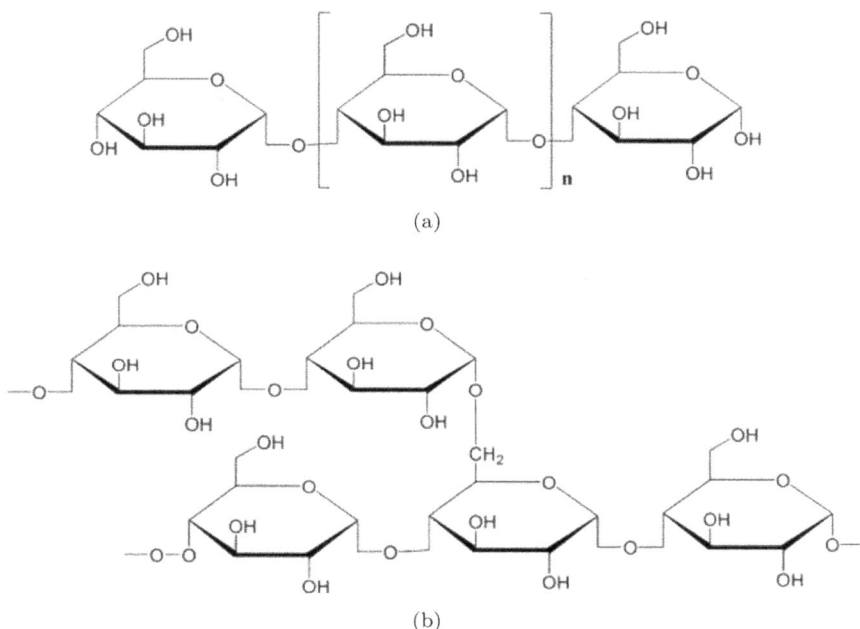

(a)

(b)

Fig. 2.4.   Structure of (a) amylose and (b) amylopectin.

Table 2.1.   Composition, granule diameter, and crystallinity of starch varying with different sources.[9]

Source	Amylose (%)	Amylopectin (%)	Granule diameter (μm)	Crystallinity (%)
Wheat	26–27	72–73	25	36
Maize	26–28	71–73	15	39
Waxy starch	<1	99	15	39
Amylomaize	50–80	20–50	10	19
Potato	20–25	74–79	40–100	25

ratio affects material performance of starch, including its mechanical properties, granule size, crystallinity, and biodegradability.[9,10]

The melting point (T_m) of pure starch is 220–240°C. However, starch also starts to degrade at around 220°C.[9] Thus, it is quite difficult to use starch alone in a high-temperature extrusion or injection process. In order to utilize starch for thermoplastic casting techniques, its processability needs to be improved by adding a plasticizer such as polyol. Plasticizers decrease the melting point of starch below its degradation point.[7] For this reason, thermoplastic starch (TPS, plasticized starch) is predominant in the bioplastic markets owing to its low price, biodegradability, and environmentally benign qualities. However, starch-based materials have critical problems such as high hygroscopicity and poor mechanical properties that are obstacles to broadening their range of applications. There are two ways to overcome these limitations: modification of hydroxyl group ($-OH$) into more hydrophobic functional groups and blending with other biodegradable polymers.[6] In order to improve vulnerability of TPS to moisture, many studies on acetylation of starch have been conducted.[10−14] It has been reported that the acetyl groups hinder inter- and intramolecular hydrogen bonds, and thus, the processability of starch is enhanced. Besides, resistance to moisture can be increased by substituting hydroxyl groups by more hydrophobic acetyl groups.[13] Optical clarity is also increased due to inhibition of the association of amylopectin, which can generate cloudiness in starch dispersions.[15] Although sensitivity to moisture reduces by acetylation of TPS, chemical treatment causes deterioration of its mechanical properties.[16] Thus, blending with biodegradable polymers such as polylactic acid (PLA),[17−20] poly(ε-caprolactone) (PCL),[21−24] and poly(vinyl alcohol) (PVA)[25−28] is needed to overcome the drawback. In the blending system, the major issue is an interfacial interaction between hydrophilic starches and hydrophobic biodegradable polymers. The key to improve compatibility is to employ plasticizers such as glycerol, formamide, and sorbitol.[29]

2.2. *Cellulose*

Cellulose is the most abundant biopolymer on earth and is a linear homopolymer composed of β-D-glucopyranose (β-D-glucose) units linked by ($1{\rightarrow}4$) glycosidic bonds unlike starch (Fig. 2.5). Cellulose is the main component of cell walls in plants and even exists

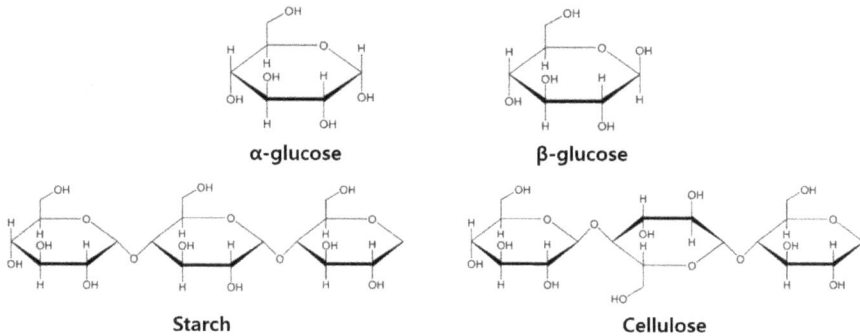

Fig. 2.5.   Structure of cellulose compared to starch.

in bacteria, fungi, algae, and some marine animals (tunicates).[30] Cellulose, extracted from plant tissues by chemical treatments, was first discovered and its molecular formula determined by the French chemist Anselme Payen in 1838. He observed that cellulose is isomeric with starch.[31]

Intra- and intermolecular hydrogen bonds appear in cellulose. They lead to a linear configuration of the chains and parallel stacking of the multiple chains. The robust hydrogen bonding interactions make cellulose extremely stable, insoluble in solvents, and highly stiff in the axial direction.[32] Also, the unique molecular structure of cellulose endows it with hydrophilicity, biodegradability, and hierarchical organization.[31]

As shown in Fig. 2.6, it exhibits a hierarchical structure ranging from a macro- to nanoscale. Cellulose fibrils are separated into a highly ordered crystalline domain and a disordered amorphous domain.[32] Crystalline cellulose possesses six polymorphs (I, II, III_I, III_{II}, IV_I, and IV_{II}), which can be interconverted as shown in Fig. 2.7.[30] Cellulose I, also referred to as native cellulose, is a slender rod-like crystalline microfibril produced in various organisms such as plants, tunicates, algae, and bacteria. Cellulose I is thermochemically metastable and can be transformed into the most stable cellulose II by regeneration (solubilization and re-precipitation) or mercerization (swelling of cellulose fibrils by treatments of aqueous

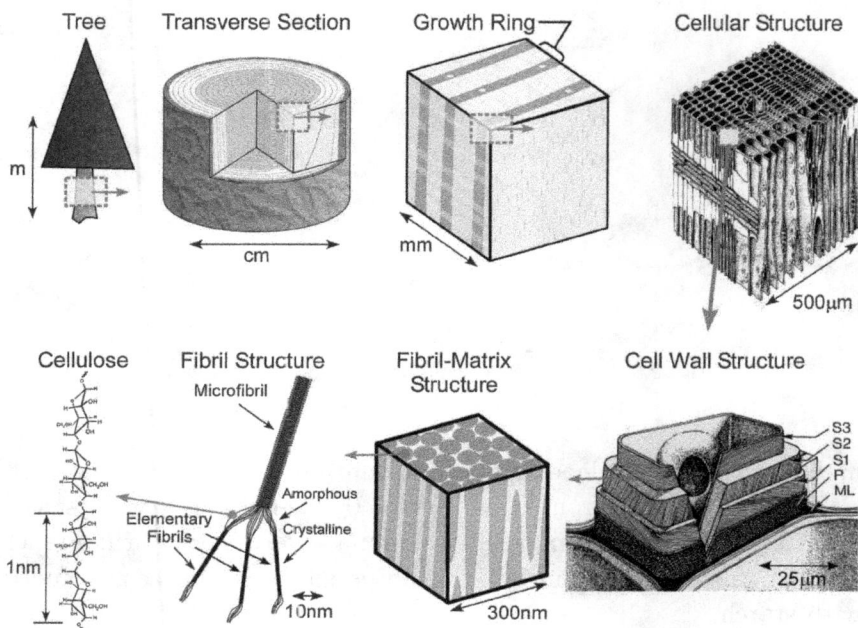

Fig. 2.6. Schematic of the hierarchical structure in a tree (cellulose).[32] (Reprinted with permission from Ref. 32. Copyright 2011 The Royal Society of Chemistry).

sodium hydroxide).[33] Cellulose III_I and III_{II} can be prepared by liquid ammonia treatments of cellulose I and II, respectively. Cellulose IV_I and IV_{II} can also be formed by heating of Cellulose III_I and III_{II}, respectively.[30]

Through CP/MAS ^{13}C NMR spectroscopic analysis and electron diffraction, it was discovered that cellulose I has two polymorphs: I_α with one-chain triclinic structure and I_β with two-chain monoclinic structure.[36] The ratio of I_α and I_β varies with the sources of cellulose. For instance, I_α phase predominates in most algae and bacteria, whereas I_β phase predominates in the cell wall of plants.[32,37,38] I_α phase, a metastable state, can be converted into the more stable I_β phase by annealing under saturated steam.[38]

Cellulose has many advantages such as biodegradability, biocompatibility, renewability and high specific strength. Thus, it

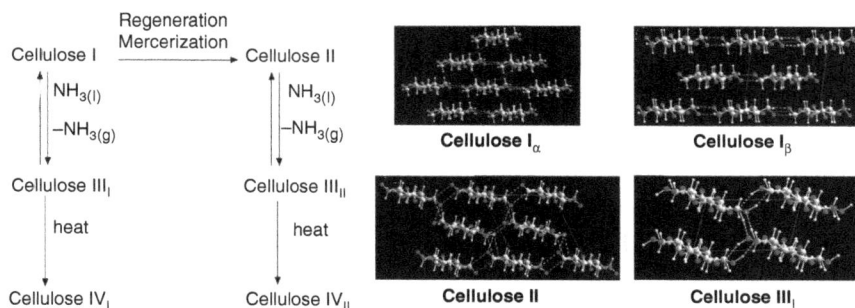

Fig. 2.7. Interconversion of crystalline cellulose polymorphs.[35] (Reprinted with permission from Ref. 35. Copyright 2004 American Chemical Society).

is extensively used not only in barrier films, antimicrobial films, and transparent films but also in flexible displays, reinforcing fillers for polymers, biomedical implants, drug delivery, batteries, supercapacitors, and separation membranes. In order to utilize cellulose for various applications, cellulose should be modified to improve its processability because of its infusibility in most solvents.[6,39] Cellulose derivatives involve cellulose ethers such as carboxymethyl cellulose (CMC), hydroxypropyl cellulose (HPC), and cellulose esters such as cellulose acetate (CA) and cellulose acetate butyrate (CAB).[40] CA is one of the most useful cellulose derivatives and finds wide applications such as coating, in photographic films, filters, pharmaceuticals, and in building construction. Performance of CA can be improved by grafting it with other polymers.[41] For better understanding of roles of cellulose in composites, it is worth noting the physical and chemical properties of cellulose, discussed in Sections 2.2.1 and 2.2.2.

In addition to cellulose derivatives (ethers and esters), cellulose is converted into regenerated materials (fibers, films, membranes and sponges) by large-scale industrial process, starting from dissolution of pulp as a feedstock. The oldest technology for producing regenerated cellulose is the viscose method in which pulp with CS_2 is transformed into cellulose xanthogenate as a metastable intermediate. The xanthogenate dissolved in aqueous sodium hydroxide can be a viscose solution in a wet process. After precipitation of the formed product, the substituent is cleaved off and then cellulose with high purity is regenerated.[31] However, the viscose process generates several environmentally hazardous by-products, such as CS_2, H_2S, and heavy metals. To minimize noxious by-products, Johnson developed a solvent system using cyclic amine oxides, particularly *N*-methylmorpholine-*N*-oxide (NMMO). NMMO can dissolve not only cellulose but also other polymers because of its strong N–O dipole.[42] The NMMO process (or the Lyocell process) reduces the processing steps compared to the viscose process as shown in Fig. 2.8. Although it is a relatively simple and resource-preserving route, the NMMO process has difficulty in adjusting the properties of the Lyocell fibers in a broad range. Lyocell fibers not only have high strength with good stability but also high fibrillation tendency in the wet state.[43] These properties can result from a high degree of crystallinity and a high orientation of the cellulose chains. The NMMO technology is being further optimized, particularly with the

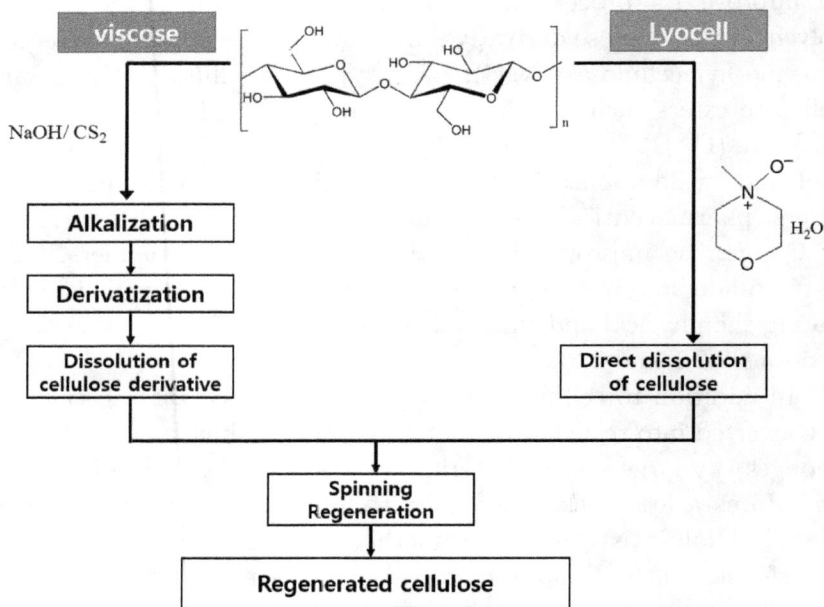

Fig. 2.8.   Comparison between viscose and Lyocell technology.

aim of decreasing fibrillation and manufacturing viscose-like fibers with good washing stability.[44]

2.2.1. *Physical and Chemical Properties of Cellulose*

The shapes of cellulose are similar to rod-like fibers, but its aspect ratio (length (L)/diameter (D)) varies depending on the source and extraction methods[45] (Table 2.2). Because cellulose elements are mostly fibrous forms, nanocellulose is suitable for reinforcement to enable stress transfer from the matrix to the filler phase.[46] Degree of polymerization (DP) is correlated with aspect ratio or length of cellulose. It is found that strength gets lowered with decreasing DP of cellulose. Thus, the DP becomes a tool to evaluate the performance of cellulose as reinforcing agents in various matrices.[47]

Even a single strand of cellulose consists of crystalline and amorphous parts. During biosynthesis, van der Waals interaction and intermolecular hydrogen bonds induce parallel stacking of multiple cellulose chains and thus generate cellulose to arrange in a highly ordered state.[45] Degrees of crystallinity mainly depend on the sources and the treatment method. Among various cellulose sources, tunicates have highest crystallinity of 85–100%, compared to wood and plants, whose crystallinity is roughly 43–65%.[48] It is not yet very clear what the true modulus of a cellulose crystal is. It is also difficult to isolate a single strand of cellulose crystal from cellulose sources because of its extremely small diameter. Sakurada *et al.* reported that

Table 2.2. Geometrical dimensions of cellulose from various sources.[48]

Source	Length (nm)	Diameter (nm)
Soft wood pulp	100–150	4–5
Hard wood pulp	140–150	4–5
Ramie	150–250	6–8
Sisal	100–500	3–5
Tunicate	1160	16
Valonia	>1000	10–20
Bacterial	100–1000	10–50

Table 2.3. Comparison of modulus between cellulose and engineering materials.

Material	Modulus (GPa)	Density (g cm^{-3})	Specific modulus (GPa g^{-1} cm^3)	Reference
Crystalline cellulose	110–220	1.6	69–137.5	32
Carbon fiber	150–500	1.8	83–278	32
Aluminum	69	2.7	26	46
Steel	200	7.8	26	46
Glass	69	2.5	28	46
Kevlar	130	1.4	93	32

the Young's modulus of cellulose was 138 GPa, which was measured by X-ray diffraction. Through these measurements, Young's moduli of cellulose crystals was found to vary in the range of 100–160 GPa.[46] Considering the low density, specific stiffness of cellulose is significantly higher than conventional structural materials as shown in Table 2.3.

Thermal properties of cellulose crystals depend on their sources, preparation, and surface modification. Generally, thermal degradation temperature (T_d) of cellulose starts at around 200–300°.[45] Roman *et al.* reported that sulfate ester group formed during sulfuric acid hydrolysis resulted in a significant reduction of T_d.[49] The coefficient of thermal expansion (CTE) of crystalline cellulose in an axial direction is approximately 0.1 ppm K^{-1}, which is lower than all plastics, most metals, and even ceramics. This property is advantageous because CTE of a polymer can be further lowered by using cellulose as a filler, such as for the application of flexible displays. As shown in Fig. 2.9, CTE of cellulose matrixes rises with increasing temperatures, whereas CTE of all-cellulose composites shows no change at around 0.1 ppm K^{-1}. This value is much lower than CTE of metals, ranging from 5 to 20 ppm K^{-1}.[50]

All asymmetric rod-like or platelet-like particles spontaneously exhibit ordered structures and liquid crystalline behaviors, which lead to nematic phase under certain conditions at sufficiently high concentrations. Thus, suspension of rod-like nanocellulose in an aqueous solution shows chiral nematic behavior, which is helical

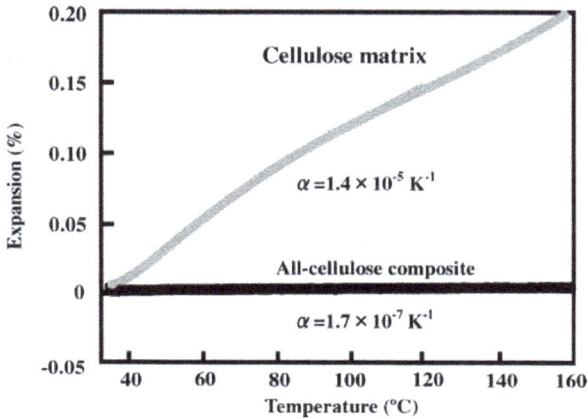

Fig. 2.9. Linear thermal expansion of cellulose matrix and all-cellulose composite.[50] (Reprinted with permission from Ref. 50. Copyright 2004 American Chemical Society).

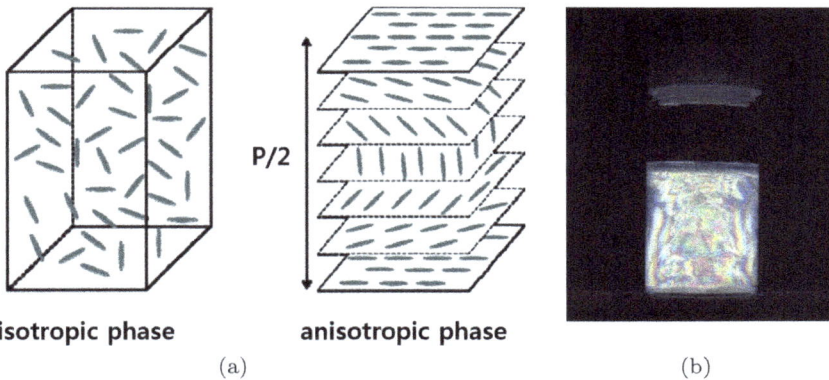

isotropic phase · anisotropic phase

(a) (b)

Fig. 2.10. (a) Schematic of orientation in nematic phase[53] (Reprinted with permission from Ref. 53. Copyright 2010 American Chemical Society), (b) birefringence of cellulose suspension immediately after shearing[54]. (Reprinted with permission from Ref. 54. Copyright 2006 American Chemical Society).

twist perpendicular to the main axis of the rod (Fig. 2.10(a)).[45] The liquid crystallinity and pitch of cellulose suspension are affected by several factors including shape, aspect ratio, dispersity, charge, electrolyte concentration, and external stimuli. It is found that liquid crystallinity of cellulose coupled to birefringence is shown Fig. 2.10(b).[51] The perceived color of cellulose films depends on the chiral nematic pitch of cholesteric order and incidence angle of light. The chiral nematic pitch can be adjusted by controlling the electrolyte concentration in cellulose suspensions. The films with this unique optical property draw attention to applications in security papers such as bank notes, ID cards, and passport, decorative films, and pigments.[52] Figure 2.10 shows the schematic of orientation in nematic phase.

Cellulose nanocrystals (CNCs) display piezoelectric property due to asymmetric crystalline structures. An electromechanical coupling effect in wood was first discovered by Bazhenov in 1950. In 1995, Fukada experimentally identified the piezoelectric coefficients of wood and demonstrated that oriented cellulose crystallites led to shear piezoelectricity.[55] The piezoelectricity of cellulose is determined by diverse factors such as density, percentage of latewood, crystallinity, moisture, and fibril orientation. The piezoelectric constant of regenerated nanocrystalline cellulose (II) was measured to be $35-60$ pC/N, which was regarded suitable for energy harvesting and power generation.[56] Thus, cellulose has potential as a soft electroactive material.

2.2.2. *Applications of Cellulose*

Researches on the applications of cellulose have been performed by considering cellulose as reinforcement fillers into polymer matrix in order to improve mechanical properties of composites by transferring stress from the matrix phase to the filler phase. There are three factors influencing reinforcing efficiency of cellulose, namely, aspect ratio, processing techniques, and compatibility. As aspect ratio increases, mechanical properties are improved because stresses are effectively transferred into entangled reinforcing fillers. However, this high aspect ratio also makes cellulose crystals difficult to be uniformly dispersed because of aggregations that prevent fillers to be efficiently reinforced.[57] It was also reported that processing techniques have an impact on the mechanical properties of a cellulose whisker-filled latex. Strengthening effectiveness decreases in the order of evaporation, hot-pressing, and extrusion. The composites prepared by extrusion or hot-pressing have lower mechanical properties than the composites prepared by evaporation, due to breakage of whiskers during the process.[58]

In general, interfacial stability of the cellulose composites is poor because polymer matrix is mostly hydrophobic in contrast to the hydrophilic cellulose fillers. External stress cannot be sufficiently transferred from the matrix to the reinforcing agents. Hence, strategies to improve compatibility and interfacial adhesion are needed. One of the strategies is to use coupling agents containing two reactive groups. One group reacts with hydroxyl groups of cellulose surface. The other group copolymerizes with the matrix and then forms covalent bonds between matrix and filler, which improves mechanical properties.[59] Network formation between cellulose fibrils also enhances mechanical properties. For instance, the mechanical property of HPC composites was not affected by DP but influenced by the quality of fibrillation. Tensile strength (TS) and elastic modulus (EM) of HPC films with cellulose networks increased by 2.5 and 4 times, respectively.[50]

The development of optoelectronic technology led to flexible and transparent displays. Among these display technologies, organic light-emitting diodes (OLEDs) have received much attention due to their attractive characteristics for display applications. Currently, commercially available OLEDs are manufactured on glass substrates, which are brittle and difficult to bend. Thus, flexible polymers are considered as alternative substrates for glass. However, there are several limitations for choosing polymer substrates, such as high CTE (exceeding 200 ppm K^{-1}), and thus, mismatch of CTE between laminates results in cracks during thermal processes. In order to overcome these limitations, nanocellulose with extremely low CTE is proposed as a substrate for flexible displays because cellulose not only reduces CTE mismatch but is also bendable and transparent.[60] Masaya Nogi *et al.* reported that bacterial-cellulose (BC)–acrylic resin composites have ultra-low-CTE of 4 ppm K^{-1},

Fig. 2.11. Flexibility of bacterial-cellulose and acrylic resin composites.[63] (Reprinted with permission from Ref. 63. Copyright 2005 American Chemical Society).

with transparency as well as flexibility (Fig. 2.11).[61] In addition, they reported that hygroscopicity, the drawback of the BC composites, was reduced through acetylation, thus maintaining transparency and thermal stability.[62] Okahisa *et al.* fabricated OLEDs successfully on a flexible, low CTE, and optically transparent cellulose nanocomposite (Fig. 2.12). The CTE of nanocomposites with wood-cellulose and ABPE could be lowered by 94%, decreasing from 213 to 12 ppm K^{-1}.[60]

Cellulose has great potential for use in biomedical and pharmaceutical fields because of its inherent biocompatibility, biodegradability, and low toxicity. Roman *et al.* conducted toxicity tests of cellulose and reported that cellulose is not toxic to animal cells at all.[64] Among cellulose sources, BC has been exclusively exploited for medical applications such as wound dressing, implants, and drug delivery. Burns are very complex injuries, leading to extensive damage to skin tissues. BC-based wound dressings such as XCell, Bioprocess, and Biofill are already commercialized. The major aim of burn therapy is to quickly achieve wound closure against the external environment

Fig. 2.12. Luminescence of OLED deposited onto a wood-cellulose nanocomposite.[60] (Reprinted with permission from Ref. 60. Copyright 2009 American Chemical Society).

in order to improve the rate of healing and relieve immediate pain. Additionally, infection and dehydration are prevented.[65] Czaja *et al.* used practical BC membranes to cure patients with severe second-degree burns, as shown in Fig. 2.13. This study reported that the skin of patients covered with never-dried cellulose membranes healed faster than those covered with conventional wet gauze.[66]

Cellulose has numerous derivatives. This versatility enables the control of the rate of drug release and the prevention of crystallization. Cellulose-based materials in drug deliveries were approved by the US Food and Drug Administration. For example, cellulose acetate has been used successfully in HIV drugs, pain relievers, antibiotics, and flavonoids.[67] Also, hydroxypropyl methylcellulose (HPMC) has been used in oral drug delivery. HPMC promoted contact with the intestinal epithelium membrane for poorly water-soluble drugs. HPMC prolonged the residence time of the drug and increased drug transportation.[68]

2.3. *Chitin/Chitosan*

Chitin, poly(β-(1→4)-*N*-acetyl-D-glucosamine) or polymer of β-(1→4)-linked 2-acetamido-2-deoxy-D-glucopyranose(*N*-acetyl-2-amino-2-

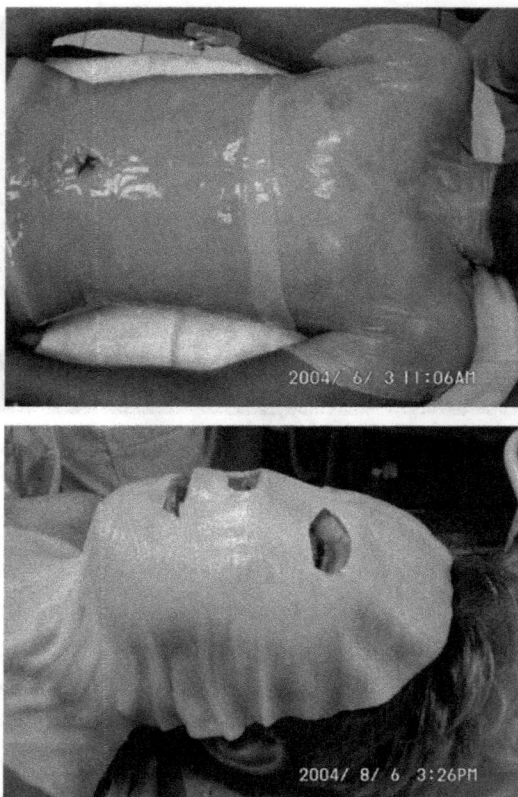

Fig. 2.13.   Application of BC in wound dressing.[66] (Reprinted with permission from Ref. 66. Copyright 2005 American Chemical Society).

deoxy-D-glucopyranose), which is an amino-polysaccharide first identified in 1884, is the second most abundant natural polymer after cellulose. In nature, chitin is one component of the exoskeleton of crustaceans and exists in ordered crystalline microfibril form. Chitin is extracted from arthropod shells such as from shrimps, crabs, and lobsters. Moreover, it can be found in wings of insects like butterflies as well as in yeasts, mushrooms, and cell wall of fungi. The structure of chitin is similar to that of cellulose, except that the acetylamino group ($-NHCOCH_3$) is located in the C2 position instead of hydroxyl group.[69-71] Chitin is an intractable material because of its

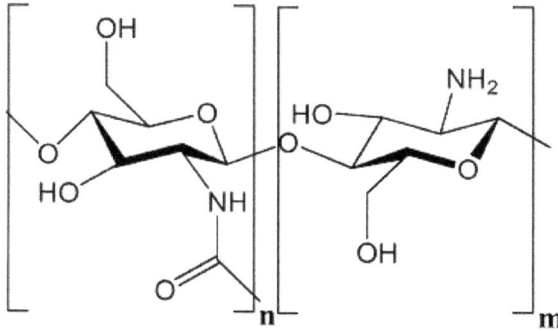

Fig. 2.14. Structure of chitin ($n \gg m$) and chitosan ($m \gg n$).

infusibility arising from strong inter- and intramolecular hydrogen bonds. Hence, chemical modification of chitin is essential to develop its processing and use. Chitosan, one of the chitin derivatives, can be obtained by *N*-acetylation of chitin. It can be prepared in various forms like gels, beads, films, sponges, and fibers. Chitosan is a weak base and is insoluble in neutral and alkaline solutions bearing deacetylated amino groups ($-NH_2$).[70] But, it is soluble in dilute aqueous acidic solutions below its pK_a (6.3) due to the presence of positive charges ($-NH_3^+$).[72] Chitin and chitosan are heteropolymers composed of monomeric units of *N*-acetyl-D-glucosamine and D-glucosamine linked by glycosidic bonds. Chitin and chitosan contain more *N*-acetyl-D-glucosamin and D-glucosamine, respectively (Fig. 2.14).[73]

Chitosan has many attractive characteristics such as biocompatibility, biodegradability, low toxicity, and biological activity. In particular, it is a suitable material in biomedical applications in that it has low immunogenicity and hemostatic, bacteriostatic, and antitumor properties.[70,74] However, chitosan-based composites have shortcomings such as sensitivity to water and low mechanical properties. The two methods to solve the problems are introduction of nanofillers and blending with other polymers. By adding nanofillers such as cellulose nanofibers[75,76] and nanoclays[77,78] into the chitosan matrix, the barrier property as well as mechanical properties were improved due to strong interfacial interactions between the matrix and the filler, causing efficient transfer of stress.[79] Another way is to blend with other polymers like PVA.[80,81] Because of interaction, mechanical properties were increased. However, chitosan composites blended with PLA exhibited decrease in strength due to incompatibility between them.[82]

3. Biopolymers

3.1. *Soy Protein*

Recently, soybean, another biomass, has been developed for use in non-food material applications such as commodities, food packaging, housing, and adhesives.[83,84] Soybean in a dry state contains around 42% protein, 33% carbohydrates, 20% oil, and 5% ash. Among these, both protein and oil can be used to make plastics and resins.[85] Soy protein comprises complex macromolecules composed of three types of amino acids, namely, acidic amino acids (aspartic acid, glutamic acid) and their amides (asparagine, glutamine), basic amino acids (lysine, arginine), and uncharged polar amino acids (glycine).[86] The soy protein product is classified into three different forms based on protein content: soy flour (SF, 54% protein), soy protein concentrate (SPC, 65–72% protein), and soy protein isolate (SPI, $\geq$ 90% protein). SF is produced by grinding raw soybeans. SPC and SPI are purified forms of SF that are produced by removing non-protein components.[84,87] Solubility of soy protein in water is highly susceptible to pH. This unique characteristic is used to extract soy proteins from raw soybean. That is, SPI is prepared by dissolving protein in raw seeds into neutral and alkaline solutions, followed by increasing acidity in order to precipitate protein. The protein has an isoelectric point at pH 4.5, at which is has minimum solubility.[88]

Soy protein-based materials have advantages of being biodegradable, environmentally friendly, and rich in resource.[85] However, applications of soy protein-based plastics are limited by low mechanical properties and high moisture absorption.[1] These drawbacks of soy protein-based plastics can be resolved by using extrusion processing with addition of plasticizers, or by adding cross-linking agents, or by mixing with other biodegradable polymers.[89] Because soy protein has many reactive sites like $-NH_2$, $-OH$, and $-SH$, it is prone to react with cross-linkers or plasticizers.[90]

The beginning of soy-based plastics as non-food applications goes back to the 1930s. At that time, Henry Ford tried to use soy protein as feedstock for plastics. He produced soy-based plastics, by mixing soy protein with phenol–formaldehyde resin, which were used as body parts of automobiles.[86] In 1939, Brother and Mckinney reported that water absorption of soy-based plastics can be reduced by adding various cross-linking agents.[91] Also, it was reported that their water sensitivity can be decreased by blending with a more hydrophobic polyester.[92] In addition, mechanical properties of the plastics can be increased by adding fillers into SPI thermoplastics such as chitin whiskers (SPI: TS = 3 MPa, EM = 20 MPa, SPI with 20 wt.% chitin whiskers: TS = 8 MPa, EM = ~160 MPa),[93] cellulose whiskers (SPI: TS = 15 MPa, EM = 500 MPa, SPI with 20 wt.% cellulose whiskers: TS = 30 MPa, EM = 1 GPa)[94] because of efficient interactions between proteins and fillers.

3.2. *Polyhydroxyalkanoates (PHAs)*

Polyhydroxyalkanoates (PHAs) are naturally occurring polyesters made by bacteria and stored as carbon and energy sources within its cytoplasm.[95] The first PHA, poly(3-hydroxybutyrate) (P3HB), was discovered by a French microbiologist, Lemoigne, in 1926. He reported that the gram-positive bacterium, *Bacillus megaterium,* synthesized homopolymers consisting of 3-hydroxybutyric monomer units in an intracellular granule.[96,97] Today, it is found that PHAs have more than 150 different types of monomers and polymers such as homopolymers, random copolymers and block copolymers, varying by microorganism species and growth conditions.[95] Figure 2.15 shows the general structure of PHAs.

PHAs are now essential materials for biomedical applications such as sutures, nerve cuffs, skin substitutes, staples, pericardial patches, cell scaffolds, and drug delivery because they are biodegradable, biocompatible, and non-toxic.[97] All PHAs are in the *R* due to

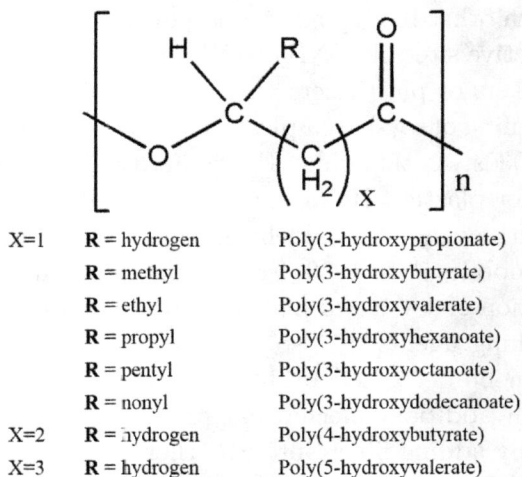

X=1	R = hydrogen	Poly(3-hydroxypropionate)
	R = methyl	Poly(3-hydroxybutyrate)
	R = ethyl	Poly(3-hydroxyvalerate)
	R = propyl	Poly(3-hydroxyhexanoate)
	R = pentyl	Poly(3-hydroxyoctanoate)
	R = nonyl	Poly(3-hydroxydodecanoate)
X=2	R = hydrogen	Poly(4-hydroxybutyrate)
X=3	R = hydrogen	Poly(5-hydroxyvalerate)

Fig. 2.15. General structure and examples of PHAs.

Table 2.4. Thermal and mechanical properties of PHAs.[95,98]

PHAs	Melting point (°C)	Tensile strength (MPa)	Young's modulus (GPa)	Elongation to break (%)
P(3HB)	175	40	3.5	5–6
P(3HV)	106.2	6.6	0.39	3–5
P(4HB)	60	50	0.07	1000
P(3HB-co-3HV)	131–170	20–25	0.7–2.9	50
P(3HB-co-4HB)	152	26	—	444
P(3HHx-co-3HO)	61	9	0.008	380
P(3HB-co-3HHx)	97	4.5	0.135	107.7

stereospecificity of enzyme used for bacteria to polymerize PHAs.[98] Physical properties of PHAs are affected by the number of carbons in the monomer unit, types of monomer, and molecular weight (Mw) as shown in Table 2.4.[95] For example, PHAs with short chain length (3–5 carbons in monomer) like P(3HB) exhibited high crystallinity, brittleness, and excellent mechanical properties. Stiffness and TSs of P(3HB) are similar to those of propylene (PP). On the other

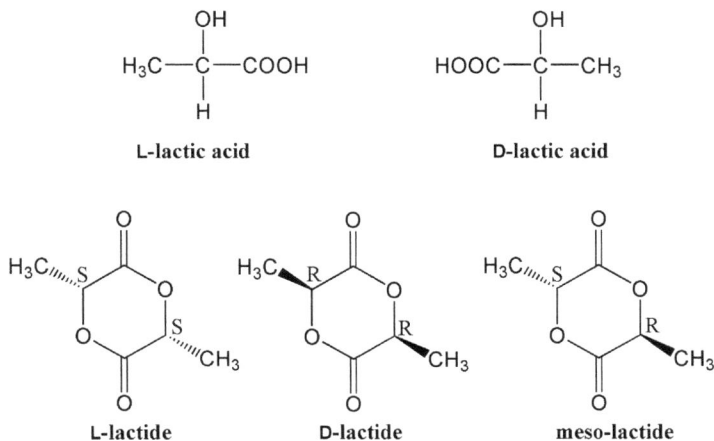

Fig. 2.16. Stereoisomers of lactic acid and lactide.

hand, PHAs containing more than one kind of monomer, namely copolymers, like poly(3-hydroxybutyrate-*co*-3-hydroxyhexanoate) (P(3HB-*co*-3HHx)) have low crystallinity, low melting temperature, ductility, and similar mechanical properties as those of low-density polyethylene (LDPE).[99] Hence, PHAs, whose properties enable wide ranges of applications from thermoplastics to elastomers, become promising alternatives to the traditional petroleum-based plastics.

3.3. *Polylactic Acid (PLA)*

Polylactic acid or polylactide (PLA) is a biodegradable aliphatic polyester made from lactic acids, which are usually produced by microbial fermentation of carbohydrate-rich substances such as corn and sugar.[100] Lactic acid, 2-hydroxypropionic acid, has two isomers, namely L- and D-lactic acid, due to existence of a chiral center (Fig. 2.16).[101] Lactic acid is produced by fermentation processes because of the high production yield of a desired stereoisomer as well as economic and environmental benefits.[102]

PLA is synthesized by direct condensation of lactic acid or ring-opening polymerization (ROP) of lactide (cyclic diester of

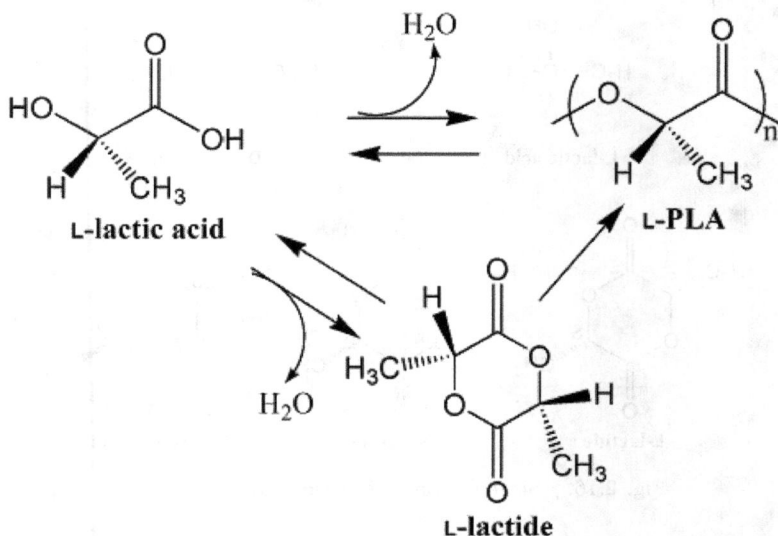

Fig. 2.17.   Two ways to polymerize lactic acids.

lactic acid) (Fig. 2.17).[103] Because lactide has three stereochemical forms, namely L-lactide with S,S-stereocenter, D-lactide with R,R-stereocenter, and meso-lactide with S,R-stereocenter, PLA exists in poly(L-lactide) (PLLA), poly(D-lactide), and poly(DL-lactide) (PDLLA) forms. Condensation polymerization is limited by achieving PLA with high Mw because of difficulty in removal of water. Hence, ROP with catalysts is preferred to produce high MW PLAs.[103]

In 1932, Carothers (DuPont) discovered PLA by heating lactic acid in vacuum. But, because of high production cost, historically, PLA had been used only for biomedical applications such as sutures.[104] However, in 1992, Cargill Dow LLC developed a low-cost continuous process for mass production of PLA. Since then, PLA has been utilized for general commodity applications. The process, developed by Cargill Dow LLC, starts with the preparation of a PLA prepolymer with low Mw by a continuous condensation, followed by converting the mixture of lactide stereoisomers with tin

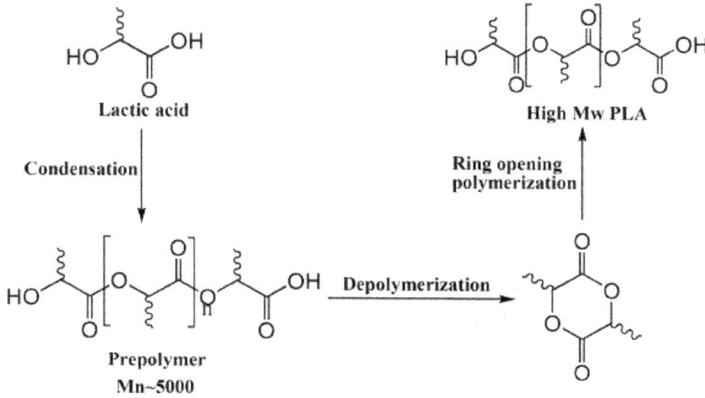

Fig. 2.18.   Schematic of low-cost PLA production developed by Cargill Dow LLC.

Table 2.5.   Properties of PLLA, PET, and PS.[106]

Material	PLLA	PET	PS
Density (kg/m^{-3})	1.26	1.40	1.05
Tensile strength (MPa)	59	57	45
Elastic modulus (GPa)	3.8	2.8–4.1	3.2
Elongation at break (%)	4–7	300	3

catalysts. PLA with high Mw is produced by ROP in the presence of tin catalysts (Fig. 2.18).[105]

PLA is a versatile thermoplastic and has many advantages such as excellent biodegradability and biocompatibility, good mechanical properties, superior oil resistance, good crease-retention, and barrier to flavors and aromas.[100] However, the processing and, thus, applications of PLA are limited because of its brittleness.[103] As shown in Table 2.5, TS and EM of PLLA are slightly higher than PET, but elongation of PET is far higher than PLLA, approximately 50 times. In other words, toughness of PLLA is significantly low compared to PET.[106] There are four ways to improve mechanical properties including toughness: modulation of stereochemical compositions,[106] orientation,[107] blending, and copolymerization.[103,108]

3.4. *Bio-polyurethane (bio-PU)*

Polyurethane (PU) is one of the most versatile polymers because it can customize its achievable properties depending on applications ranging from rigid plastics to viscoelastic gels. While the application of PU is not limited, large quantities of PU are used in furniture and automotive industries, such as for seat cushions, bumpers, and sound insulation. Conventional PU is usually synthesized from polyols and isocyanates derived from petroleum. Thus, crude oil price severely influences PU production costs.[109] With these economic and environmental issues of petroleum sources, it is important to diversify PU feedstocks to renewable resources. So, many researchers have developed processes for producing bio-PU (bio-based PU) by using biopolyols and bioisocyanates, directly derived from plant oils.[110] Plant oils are usually triglycerides, which are ester compounds of one glycerol and three fatty acids. Figure 2.19 shows structures of several fatty acids. Among them, only ricinoleic acid has inherent hydroxyl groups. The composition of fatty acids in a few plant oils are summarized in Table 2.6.[111]

Except castor oil which contains ricinoleic acid, most plant oils do not possess natural hydroxyl groups. Because hydroxyl groups are essentially needed to react with isocyanate, double bonds of the fatty acids are converted into hydroxyl groups.[111] There are three

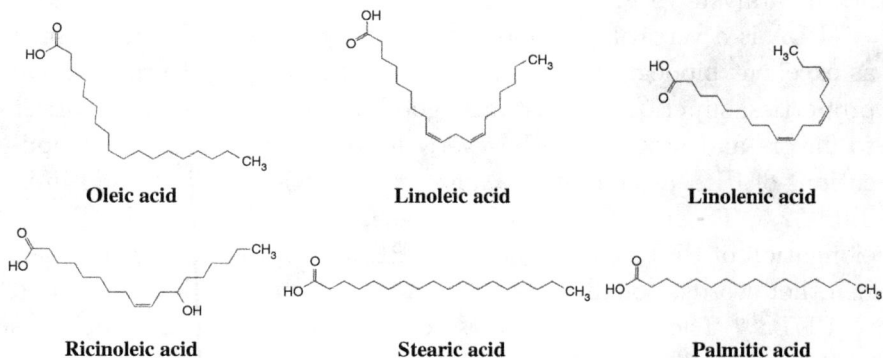

Fig. 2.19.   Structures of some fatty acids.

methods to chemically introduce hydroxyl groups into the unsaturated sites[112–114]: epoxidation followed by methanolysis (oxirane opening), polyol, hydroformylation followed by hydrogenation, polyol II, and ozonolysis followed by hydrogenation, polyol III (Fig. 2.20). While polyol I has secondary hydroxyl groups, polyol II has primary hydroxyl groups. Thus, polyol II has higher PU production yield than

Table 2.6. Composition of fatty acids in some plant oils.[111]

| Fatty acid (%) | Plant oil | | | |
	Castro oil	Soybean oil	Palm oil	Sunflower oil
Palmitic acid	1.5	12	39	6
Stearic acid	0.5	4	5	4
Oleic acid	5	24	45	42
Linoleic acid	4	53	9	47
Linolenic acid	0.5	7	—	1
Ricinoleic acid	87.5	—	—	—
Others	—	—	2	—

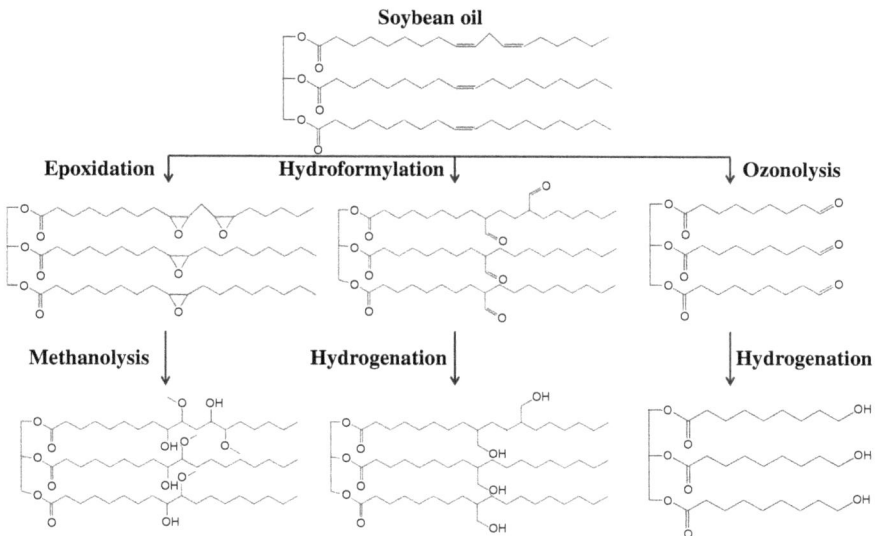

Fig. 2.20. Three methods for conversion of soybean oil into polyol.

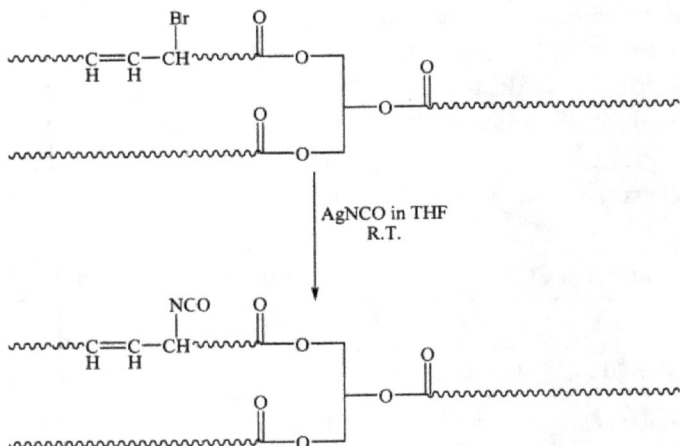

Fig. 2.21. Synthesis of isocyanate from soybean oil.[116]

polyol I because isocyanate has preferred reactivity toward primary hydroxyl groups.[113] Polyol III is more reactive with isocyanate than the other two polyols because it forms primary hydroxyl groups at the terminal of a carbon chain.[115]

Unlike polyol, little research has been conducted on derivative modification of isocyanate from plant oils. Gökhan Çaylı *et al.* reported that bioisocyanate was successfully synthesized by reacting AgNCO with soybean oil substituted by ally bromide (Fig. 2.21).[116]

4. Conclusion

This chapter presents an introduction to biopolymers. The usage of biopolymers can lead to reduction of CO_2 emissions and waste. Biopolymers are categorized into natural polymers and polymers synthesized from natural monomers. Polysaccharides, one of the most common natural polymers, consist of sugars linked by glycosidic bonds. They have several characteristics: abundance, non-toxicity, biodegradability, biocompatibility, high mechanical properties, and hygroscopic properties. Hydrophilicity and hygroscopicity are obstacles to using polysaccharides in various applications. But, this limitation can be overcome by surface modification. Among polysaccharides, celluloses are reviewed in detail. Cellulose features an attractive combination of properties such as high aspect ratio, crystallinity, high specific EM, low CTE, optical transparency, anisotropy, biodegradability, and biocompatibility. The potential applications of cellulose include fillers in composites, security papers, electronic devices, wound dressings, and drug delivery.

PHAs are useful materials in biomedical applications due to biodegradability, non-toxicity, high biocompatibility, and a wide range of mechanical properties. PLA and bio-PU are polymerized using sugars and plant oils (fatty acids), respectively. PLA has advantages such as biodegradability, biocompatibility, moderate TS, an EM comparable to petro-based polymers (PET, PS), superior oil resistance, good crease-retention, and a high gas barrier. However, processing of PLA is limited by low toughness. Toughness can be improved by modulating stereochemical composition, orientation, blending, and copolymerization. There are some efforts to substitute bio-PU for conventional PU because of cost sensitivity according to changing oil price. Recently, various methods to make bio-PU have been developed and bio-PU has been utilized in the furniture and automotive industries for seat cushions, bumpers, and sound insulation.

References

1. A. K. Mohanty, M. Misra and L. T. Drzal, Sustainable bio-composites from renewable resources: Opportunities and challenges in the green materials world, *J. Polym. Environ.* **10**(1–2), 19–26 (2002).
2. U.S. Congress, *Office of Technology Assessment, Biopolymers: Making Materials Nature's Way-Background Paper*, Washington, DC: US Government Printing Office (1993).
3. S. V. N. Vijayendra and T. R. Shamala, Film forming microbial biopolymers for commercial applications — A review, *Crit. Rev. Biotechnol.* **34**(4), 338–357 (2014).
4. V. R. Sinha and R. Kumria, Polysaccharides in colon-specific drug delivery, *Int. J. Pharm.* **224**(1–2), 19–38 (2001).
5. G. Crini, Recent developments in polysaccharide-based materials used as adsorbents in wastewater treatment, *Prog. Polym. Sci.* **30**(1), 38–70 (2005).
6. I. Vroman and L. Tighzert, Biodegradable polymers, *Mater.* **2**(2), 307–344 (2009).

7. A. C. Wu, T. Witt and R. G. Gilbert, Characterization methods for starch-based materials: State of the art and perspectives, *Austr. J. Chem.* **66**(12), 1550–1563 (2013).

8. X. L. Wang, K. K. Yang and Y. Z. Wang, Properties of starch blends with biodegradable polymers, *J. Macromol. Sci. Polymer Rev.* **C43**(3), 385–409 (2003).

9. L. Averous, Biodegradable multiphase systems based on plasticized starch: A review, *J. Macromol. Sci. Polymer Rev.* **C44**(3), 231–274 (2004).

10. C. Fringant, J. Desbrieres and M. Rinaudo, Physical properties of acetylated starch-based materials: Relation with their molecular characteristics, *Polymer* **37**(13), 2663–2673 (1996).

11. B. Volkert, A. Lehmann, T. Greco and M. H. Nejad, A comparison of different synthesis routes for starch acetates and the resulting mechanical properties, *Carbohydr. Polym.* **79**(3), 571–577 (2010).

12. P. Baruk Zamudio-Flores, A. Vargas Torres, R. Salgado-Delgado and L. Arturo Bello-Perez, Influence of the oxidation and acetylation of banana starch on the mechanical and water barrier properties of modified starch and modified starch/chitosan blend films, *J. Appl. Polym. Sci.* **115**(2), 991–998 (2010).

13. W. Jiang, X. Qiao and K. Sun, Mechanical and thermal properties of thermoplastic acetylated starch/poly(ethylene-co-vinyl alcohol) blends, *Carbohydr. Polym.* **65**(2), 139–143 (2006).

14. K. Katerinopoulou, A. Giannakas, K. Grigoriadi, N. M. Barkoula and A. Ladavos, Preparation and characterization of acetylated corn starch-(PVOH)/clay nanocomposite films, *Carbohydr. Polym.* **102**, 216–222 (2014).

15. O. S. Lawal and K. O. Adebowale, Physicochemical characteristics and thermal properties of chemically modified jack bean (*Canavalia ensiformis*) starch, *Carbohydr. Polym.* **60**(3), 331–341 (2005).

16. S. A. A. Ghavimi, M. H. Ebrahimzadeh, M. Solati-Hashjin and N. A. Abu Osman, Polycaprolactone/starch composite: Fabrication, structure, properties, and applications, *J. Biomed. Mater. Res. Part A* **103**(7), 2482–2498 (2015).

17. J. F. Zhang and X. Z. Sun, Mechanical properties of poly(lactic acid)/starch composites compatibilized by maleic anhydride, *Biomacromolecules* **5**(4), 1446–1451 (2004).

18. M. A. Huneault and H. Li, Morphology and properties of compatibilized polylactide/thermoplastic starch blends, *Polymer* **48**(1), 270–280 (2007).

19. S. Y. Lee, M. A. Hanna and D. D. Jones, An adaptive neuro-fuzzy inference system for modeling mechanical properties of tapioca starch-poly(lactic acid) nanocomposite foams, *Starch-Stärke* **60**(3–4), 159–164 (2008).

20. Y. Yu, Y. Cheng, J. Ren, E. Cao, X. Fu and W. Guo, Plasticizing effect of poly(ethylene glycol)s with different molecular weights in poly(lactic acid)/starch blends, *J. Appl. Polym. Sci.* **132**(16), 41808–41816 (2015).

21. L. Averous, L. Moro, P. Dole and C. Fringant, Properties of thermoplastic blends: starch-polycaprolactone, *Polymer* **41**(11), 4157–4167 (2000).
22. C. S. Wu, Physical properties and biodegradability of maleated-polycaprolactone/starch composite, *Polym. Degrad. Stabil.* **80**(1), 127–134 (2003).
23. R. Ortega-Toro, I. Morey, P. Talens and A. Chiralt, Active bilayer films of thermoplastic starch and polycaprolactone obtained by compression molding, *Carbohydr. Polym.* **127**, 282–290 (2015).
24. X. Zhang, Y. Zhang, J. Liao, T. Yu, R. Hu, Z. Wu and Q. Wu, Preparation and properties of compatible starch-polycaprolactone composites: Effects of molecular weight of soft segments in polyurethane compatilizer, *J. Appl. Polym. Sci.* **132**(32), 42381–42390 (2015).
25. R. Jayasekara, I. Harding, I. Bowater, G. B. Y. Christie and G. T. Lonergan, Preparation, surface modification and characterisation of solution cast starch PVA blended films, *Polym. Test.* **23**(1), 17–27 (2004).
26. H. R. Park, S. H. Chough, Y. H. Yun and S. D. Yoon, Properties of starch/PVA blend films containing citric acid as additive, *J. Polym. Environ.* **13**(4), 375–382 (2005).
27. B. Sreedhar, D. K. Chattopadhyay, M. S. H. Karunakar and A. R. K. Sastry, Thermal and surface characterization of plasticized starch polyvinyl alcohol blends crosslinked with epichlorohydrin, *J. Appl. Polym. Sci.* **101**(1), 25–34 (2006).
28. G. K. Athira and A. N. Jyothi, Cassava starch-poly(vinyl alcohol) nanocomposites for the controlled delivery of curcumin in cancer prevention and treatment, *Starch-Stärke* **67**(5–6), 549–558 (2015).
29. D. R. Lu, C. M. Xiao and S. J. Xu, Starch-based completely biodegradable polymer materials, *Exp. Polym. Lett.* **3**(6), 366–375 (2009).
30. A. C. Osullivan, Cellulose: The structure slowly unravels, *Cellulose* **4**(3), 173–207 (1997).
31. D. Klemm, B. Heublein, H. P. Fink and A. Bohn, Cellulose: Fascinating biopolymer and sustainable raw material, *Angew. Chem. Int. Ed.* **44**(22), 3358–3393 (2005).
32. R. J. Moon, A. Martini, J. Nairn, J. Simonsen and J. Youngblood, Cellulose nanomaterials review: Structure, properties and nanocomposites, *Chem. Soc. Rev.* **40**(7), 3941–3994 (2011).
33. P. Langan, Y. Nishiyama and H. Chanzy, X-ray structure of mercerized cellulose II at 1 angstrom resolution, *Biomacromolecules* **2**(2), 410–416 (2001).
34. S. Nikolov, M. Petrov, L. Lymperakis, M. Friak, C. Sachs, H. O. Fabritius, D. Raabe and J. Neugebauer, Revealing the design principles of high-performance biological composites using *ab initio* and multiscale simulations: The example of lobster cuticle, *Adv. Mater.* **22**(4), 519–526 (2010).
35. M. Wada, H. Chanzy, Y. Nishiyama and P. Langan, Cellulose IIII crystal structure and hydrogen bonding by synchrotron X-ray and neutron fiber diffraction, *Macromolecules* **37**(23), 8548–8555 (2004).

36. H. Kono, S. Yunoki, T. Shikano, M. Fujiwara, T. Erata and M. Takai, CP/MAS C-13 NMR study of cellulose and cellulose derivatives. 1. Complete assignment of the CP/MAS C-13 NMR spectrum of the native cellulose, *J. Am. Chem. Soc.* **124**(25), 7506–7511 (2002).

37. M. Samir, F. Alloin and A. Dufresne, Review of recent research into cellulosic whiskers, their properties and their application in nanocomposite field, *Biomacromolecules* **6**(2), 612–626 (2005).

38. H. Yamamoto and F. Horii, CP MAS C-13 NMR Analysis of the crystal transformation induced for valonia cellulose by annealing at high-temperatures, *Macromolecules* **26**(6), 1313–1317 (1993).

39. G. Siqueira, J. Bras and A. Dufresne, Cellulose whiskers versus microfibrils: Influence of the nature of the nanoparticle and its surface functionalization on the thermal and mechanical properties of nanocomposites, *Biomacromolecules* **10**(2), 425–432 (2009).

40. J. W. Rhim and P. K. W. Ng, Natural biopolymer-based nanocomposite films for packaging applications, *Crit. Rev. Food Sci. Nutr.* **47**(4), 411–433 (2007).

41. T. Mekonnen, P. Mussone, H. Khalil and D. Bressler, Progress in bio-based plastics and plasticizing modifications, *J. Mater. Chem. A* **1**(43), 13379–13398 (2013).

42. C. Yin, J. Li, Q. Xu, Q. Peng, Y. Liu and X. Shen, Chemical modification of cotton cellulose in supercritical carbon dioxide: Synthesis and characterization of cellulose carbamate, *Carbohydr. Polym.* **67**(2), 147–154 (2007).

43. X. Chen, C. Burger, D. Fang, D. Ruan, L. Zhang, B. S. Hsiao and B. Chu, X-ray studies of regenerated cellulose fibers wet spun from cotton linter pulp in NaOH/thiourea aqueous solutions, *Polymer* **47**(8), 2839–2848 (2006).

44. H.-P Fink, P. Weigel, H. J. Purz, and J. Ganster, Structure formation of regenerated cellulose materials from NMMO-solutions, *Prog. Polym. Sci.* **26**(9), 1473–1524 (2001).

45. N. Lin, J. Huang and A. Dufresne, Preparation, properties and applications of polysaccharide nanocrystals in advanced functional nanomaterials: A review, *Nanoscale* **4**(11), 3274–3294 (2012).

46. S. J. Eichhorn, A. Dufresne, M. Aranguren, N. E. Marcovich, J. R. Capadona, S. J. Rowan, C. Weder, W. Thielemans, M. Roman, S. Renneckar, W. Gindl, S. Veigel, J. Keckes, H. Yano, K. Abe, M. Nogi, A. N. Nakagaito, A. Mangalam, J. Simonsen, A. S. Benight, A. Bismarck, L. A. Berglund and T. Peijs, Review: Current international research into cellulose nanofibres and nanocomposites, *J. Mater. Sci.* **45**(1), 1–33 (2010).

47. T. Zimmermann, N. Bordeanu and E. Strub, Properties of nanofibrillated cellulose from different raw materials and its reinforcement potential, *Carbohydr. Polym.* **79**(4), 1086–1093 (2010).

48. S. Rebouillat and F. Pla, State of the art manufacturing and engineering of nanocellulose: A review of available data and industrial applications, *J. Biomater. Nanobiotechnol.* **04**(02), 24 (2013).

49. M. Roman and W. T. Winter, Effect of sulfate groups from sulfuric acid hydrolysis on the thermal degradation behavior of bacterial cellulose, *Biomacromolecules* **5**(5), 1671–1677 (2004).
50. T. Nishino, I. Matsuda and K. Hirao, All-cellulose composite, *Macromolecules* **37**(20), 7683–7687 (2004).
51. R. H. Marchessault, F. F. Morehead and N. M. Walter, Liquid crystal systems from fibrillar polysaccharides, *Nature* **184**, 632–633 (1959).
52. J. F. Revol, L. Godbout and D. G. Gray, Solid self-assembled films of cellulose with chiral nematic order and optically variable properties, *J. Pulp Paper Sci.* **24**(5), 146–149 (1998).
53. Y. Habibi, L. A. Lucia and O. J. Rojas, Cellulose nanocrystals: Chemistry, self-assembly, and applications, *Chem. Rev.* **110**(6), 3479–3500 (2010).
54. J. Araki and S. Kuga, Effect of trace electrolyte on liquid crystal type of cellulose microcrystals, *Langmuir* **17**(15), 4493–4496 (2001).
55. J. Kim and S. Yun, Discovery of cellulose as a smart material, *Macromolecules* **39**(12), 4202–4206 (2006).
56. L. Csoka, I. C. Hoeger, O. J. Rojas, I. Peszlen, J. J. Pawlak and P. N. Peralta, Piezoelectric effect of cellulose nanocrystals thin films, *ACS Macro Lett.* **1**(7), 867–870 (2012).
57. D. Klemm, F. Kramer, S. Moritz, T. Lindstrom, M. Ankerfors, D. Gray and A. Dorris, Nanocelluloses: A new family of nature-based materials, *Angew. Chem. Int. Ed.* **50**(24), 5438–5466 (2011).
58. P. Hajji, J. Y. Cavaille, V. Favier, C. Gauthier and G. Vigier, Tensile behavior of nanocomposites from latex and cellulose whiskers, *Polym. Compos.* **17**(4), 612–619 (1996).
59. M. Abdelmouleh, S. Boufi, M. N. Belgacem and A. Dufresne, Short natural-fibre reinforced polyethylene and natural rubber composites: Effect of silane coupling agents and fibres loading, *Compos. Sci. Technol.* **67**(7–8), 1627–1639 (2007).
60. Y. Okahisa, A. Yoshida, S. Miyaguchi and H. Yano, Optically transparent wood-cellulose nanocomposite as a base substrate for flexible organic light-emitting diode displays, *Compos. Sci. Technol.* **69**(11–12), 1958–1961 (2009).
61. M. Nogi and H. Yano, Transparent nanocomposites based on cellulose produced by bacteria offer potential innovation in the electronics device industry, *Adv. Mater.* **20**(10), 1849–1852 (2008).
62. S. Ifuku, M. Nogi, K. Abe, K. Handa, F. Nakatsubo and H. Yano, Surface modification of bacterial cellulose nanofibers for property enhancement of optically transparent composites: Dependence on acetyl-group DS, *Biomacromolecules* **8**(6), 1973–1978 (2007).
63. H. Yano, J. Sugiyama, A. N. Nakagaito, M. Nogi, T. Matsuura, M. Hikita and K. Handa, Optically transparent composites reinforced with networks of bacterial nanofibers, *Adv. Mater.* **17**(2), 153–155 (2005).
64. M. Roman, S. Dong, A. Hirani and Y. W. Lee, Cellulose nanocrystals for drug delivery, *ACS Symp. Ser.* **1017**, 81–91 (2009).

65. M. Jorfi and E. J. Foster, Recent advances in nanocellulose for biomedical applications, *J. Appl. Polym. Sci.* **132**(14), 41719–41738 (2015).

66. W. K. Czaja, D. J. Young, M. Kawecki, R. M. Brown Jr., The future prospects of microbial cellulose in biomedical applications, *Biomacromolecules* **8**(1), 1–12 (2007).

67. M. Tran and C. Wang, Semi-solid materials for controlled release drug formulation: Current status and future prospects, *Front. Chem. Sci. Eng.* **8**(2), 225–232 (2014).

68. G. A. Ilevbare, H. Liu, K. J. Edgar and L. S. Taylor, Impact of polymers on crystal growth rate of structurally diverse compounds from aqueous solution, *Mol. Pharm.* **10**(6), 2381–2393 (2013).

69. I. S. Arvanitoyannis, A. Nakayama and S. Aiba, Chitosan and gelatin based edible films: State diagrams, mechanical and permeation properties, *Carbohydr. Polym.* **37**(4), 371–382 (1998).

70. S. Rajendran and S. C. Anand, Developments in medical textiles, *Textile Prog.* **32**(4), 1–42 (2002).

71. M. Rinaudo, Chitin and chitosan: Properties and applications, *Prog. Polym. Sci.* **31**(7), 603–632 (2006).

72. R. C. Goy, D. de Britto and O. B. G. Assis, A review of the antimicrobial activity of chitosan, *Polimeros-Ciencia E Tecnologia* **19**(3), 241–247 (2009).

73. F. D. Bobbink, J. Zhang, Y. Pierson, X. Chen and N. Yan, Conversion of chitin derived *N*-acetyl-D-glucosamine (NAG) into polyols over transition metal catalysts and hydrogen in water, *Green Chem.* **17**(2), 1024–1031 (2015).

74. A. S. Sowmya, P. T. S. Kumar, S. Deepthi, K. P. Chennazhi, H. Ehrlich, M. Tsurkan and R. Jayakumar, Chitin and chitosan in selected biomedical applications, *Prog. Polym. Sci.* **39**(9), 1644–1667 (2014).

75. R. A. Khan, S. Salmieri, C. Le Tien, B. Riedl, J. Bouchard, G. Chauve, V. Tan, M. R. Kamal and M. Lacroix, Mechanical and barrier properties of nanocrystalline cellulose reinforced chitosan based nanocomposite films, *Carbohydr. Polym.* **90**(4), 1601–1608 (2012).

76. T. Wu, R. Farnood, K. O'Kelly and B. Chen, Mechanical behavior of transparent nanofibrillar cellulose-chitosan nanocomposite films in dry and wet conditions, *J. Mech. Behav. Biomed. Mater.* **32**, 279–286 (2014).

77. Y. X. Xu, X. Ren and M. A. Hanna, Chitosan/clay nanocomposite film preparation and characterization, *J. Appl. Polym. Sci.* **99**(4), 1684–1691 (2006).

78. K. Lewandowska, A. Sionkowska, B. Kaczmarek and G. Furtos, Characterization of chitosan composites with various clays, *Int. J. Biol. Macromol.* **65**, 534–541 (2014).

79. S. H. Othman, Bio-nanocomposite materials for food packaging applications: Types of biopolymer and nano-sized filler, *Agric. Agric. Sci. Procedia.* **2**, 296–303 (2014).

80. M. Li, S. Cheng and H. Yan, Preparation of crosslinked chitosan/poly(vinyl alcohol) blend beads with high mechanical strength, *Green Chem.* **9**(8), 894–898 (2007).

81. Y. Nakano, Y. Bin, M. Bando, T. Nakashima, T. Okuno, H. Kurosu and M. Matsuo, Structure and mechanical properties of chitosan/poly(vinyl alcohol) blend films, *Macromol. Sympo.* **258**(1), 63–81 (2007).

82. N. E. Suyatma, A. Copinet, L. Tighzert and V. Coma, Mechanical and barrier properties of biodegradable films made from chitosan and poly (lactic acid) blends, *J. Polym. Environ.* **12**(1), 1–6 (2004).

83. R. R. Koshy, S. K. Mary, S. Thomas and L. A. Pothan, Environment friendly green composites based on soy protein isolate — A review, *Food Hydrocoll.* **50**, 174–192 (2015).

84. P. Saenghirunwattana, A. Noomhorm, V. Rungsardthong, Mechanical properties of soy protein based "green" composites reinforced with surface modified cornhusk fiber, *Ind. Crops Prod.* **60**, 144–150 (2014).

85. J. Zhang, P. Mungara and J. Jane, Mechanical and thermal properties of extruded soy protein sheets, *Polymer* **42**(6), 2569–2578 (2001).

86. S. N. Swain, S. M. Biswal, P. K. Nanda and P. L. Nayak, Biodegradable soy-based plastics: Opportunities and challenges, *J. Polym. Environ.* **12**(1), 35–42 (2004).

87. D. Kalman, Amino acid composition of an organic brown rice protein concentrate and isolate compared to soy and whey concentrates and isolates, *Foods* **3**(3), 394 (2014).

88. R. Kumar, V. Choudhary, S. Mishra, I. K. Varma and B. Mattiason, Adhesives and plastics based on soy protein products, *Ind. Crops Prod.* **16**(3), 155–172 (2002).

89. A. K. Mohanty, P. Tummala, W. Liu, M. Misra, P. V. Mulukutla and L. T. Drzal, Injection molded biocomposites from soy protein based bioplastic and short industrial hemp fiber, *J. Polym. Environ.* **13**(3), 279–285 (2005).

90. P. C. Z. Chen and J. Jane, Biodegradable plastic made from soybean products. II. Effects of cross-linking and cellulose incorporation on mechanical properties and water absorption, *J. Environ. Polym. Degrad.* **2**(3), 211–217 (1994).

91. G. H. Brother and L. L. McKinney, Protein plastics from soybean products, *Ind. Eng. Chem.* **32**(7), 84–87 (1940).

92. P. Mungara, T. Chang, J. Zhu and J. Jane, Processing and physical properties of plastics made from soy protein polyester blends, *J. Polym. Environ.* **10**(1), 31–37 (2002).

93. Y. S. Lu, L. H. Weng and L. N. Zhang, Morphology and properties of soy protein isolate thermoplastics reinforced with chitin whiskers, *Biomacromolecules* **5**(3), 1046–1051 (2004).

94. Y. Wang, X. Cao and L. Zhang, Effects of cellulose whiskers on properties of soy protein thermoplastics, *Macromol. Biosci.* **6**(7), 524–531 (2006).

95. R. Nigmatullin, P. Thomas, B. Lukasiewicz, H. Puthussery and I. Roy, Poly-hydroxyalkanoates, a family of natural polymers, and their applications in drug delivery, *J. Chem. Technol. Biotechnol.* **90**(7), 1209–1221 (2015).

96. A. J. Anderson and E. A. Dawes, Occurrence, metabolism, metabolic role, and industrial uses of bacterial polyhydroxyalkanoates, *Microbiol. Rev.* **54**(4), 450–472 (1990).

97. M. Zinn, B. Witholt and T. Egli, Occurrence, synthesis and medical appli-cation of bacterial polyhydroxyalkanoate, *Adv. Drug Deliv. Rev.* **53**(1), 5–21 (2001).

98. K. Sudesh, H. Abe and Y. Doi, Synthesis, structure and properties of polyhy-droxyalkanoates: Biological polyesters, *Prog. Polym. Sci.* **25**(10), 1503–1555 (2000).

99. R. Rai, T. Keshavarz, J. A. Roether, A. R. Boccaccini and I. Roy, Medium chain length polyhydroxyalkanoates, promising new biomedical materials for the future, *Mater. Sci. Eng. R-Rep.* **72**(3), 29–47 (2011).

100. F. Carrasco, P. Pages, J. Gamez-Perez, O. O. Santana and M. L. Maspoch, Processing of poly(lactic acid): Characterization of chemical structure, ther-mal stability and mechanical properties, *Polym. Degrad. Stabil.* **95**(2), 116–125 (2010).

101. M. G. Adsul, A. J. Varma and D. V. Gokhale, Lactic acid production from waste sugarcane bagasse derived cellulose, *Green Chem.* **9**(1), 58–62 (2007).

102. R. P. John, K. M. Nampoothiri and A. Pandey, Fermentative production of lactic acid from biomass: An overview on process developments and future perspectives, *Appl. Microbiol. Biotechnol.* **74**(3), 524–534 (2007).

103. K. M. Nampoothiri, N. R. Nair and R. P. John, An overview of the recent developments in polylactide (PLA) research, *Bioresource Technol.* **101**(22), 8493–8501 (2010).

104. M. Flieger, M. Kantorova, A. Prell, T. Rezanka and J. Votruba, Biodegrad-able plastics from renewable sources, *Folia Microbiologica* **48**(1), 27–44 (2003).

105. R. E. Drumright, P. R. Gruber and D. E. Henton, Polylactic acid technology, *Adv. Mater.* **12**(23), 1841–1846 (2000).

106. K. S. Anderson, K. M. Schreck and M. A. Hillmyer, Toughening polylactide, *Polym. Rev.* **48**(1), 85–108 (2008).

107. R. L. Shogren, G. Selling and J. L. Willett, Effect of orientation on the morphology and mechanical properties of PLA/starch composite filaments, *J. Polym. Environ.* **19**(2), 329–334 (2011).

108. R. Z. Xiao, Z. W. Zeng, G. L. Zhou, J. J. Wang, F. Z. Li and A. M. Wang, Recent advances in PEG-PLA block copolymer nanoparticles, *Int. J. Nanomed.* **5**, 1057–1065 (2010).

109. P. Cinelli, I. Anguillesi and A. Lazzeri, Green synthesis of flexible polyurethane foams from liquefied lignin, *Eur. Polym. J.* **49**(6), 1174–1184 (2013).

110. C. Zhang, S. A. Madbouly and M. R. Kessler, Biobased polyurethanes prepared from different vegetable oils, *ACS Appl. Mater. Interfaces* **7**(2), 1226–1233 (2015).

111. F. S. Guner, Y. Yagci and A. T. Erciyes, Polymers from triglyceride oils, *Prog. Polym. Sci.* **31**(7), 633–670 (2006).

112. R. Gu, M. M. Sain and S. K. Konar, A feasibility study of polyurethane composite foam with added hardwood pulp, *Ind. Crops Prod.* **42**, 273–279 (2013).

113. G. W. Zhang and Z. S. Petrovic, Structure-property relationships in polyurethanes derived from soybean oil, *J. Mater. Sci.* **41**(15), 4914–4920 (2006).

114. L. Zhang, H. K. Jeon, J. Malsam, R. Herrington and C. W. Macosko, Substituting soybean oil-based polyol into polyurethane flexible foams, *Polymer* **48**(22), 6656–6667 (2007).

115. Z. S. Petrovic, W. Zhang and I. Javni, Structure and properties of polyurethanes prepared from triglyceride polyols by ozonolysis, *Biomacromolecules* **6**(2), 713–719 (2005).

116. G. Cayli and S. Kuesefoglu, Biobased polyisocyanates from plant oil triglycerides: Synthesis, polymerization, and characterization, *J. Appl. Polym. Sci.* **109**(5), 2948–2955 (2008).

Chapter 3

Sensing Principles

Young-Jun Lee and Joo-Hyung Kim*

Laboratory of Intelligent Devices and Thermal Control
Department of Mechanical Engineering, Inha University
100 Inha-Ro, Nam-Ku, Incheon 22212, Republic of Korea
*joohyung.kim@inha.ac.kr

Sensor is a device to convert physical world to the relevant output signal to be recognized. As an interface between them, sensor can help us to see, hear, smell and taste some physical, chemical, optical responses from its environments. Most of sensors in use communicate with an electronic system. This chapter includes some basic sensor technologies including sensor electronics and various types of sensors.

1. Basic Sensor Technologies

A sensor is a device that can convert a non-electrical physical or chemical quantity into an electrical signal.[1] This signal is produced by receiving some form of energy, such as heat, light, motion or through a chemical reaction. Therefore, a sensor acts as an interface between the real world and the world of electronic or mechanical devices to give corresponding feedback to users or other devices. Once a sensor detects one or more of these signals (an input), it converts them into an analog or digital representation of the input signal (Fig. 3.1). Most of the sensors in use communicate with an electric device that is measuring and recording. Today, different forms of sensors are employed for a wide range of functions (i.e., the touchscreen

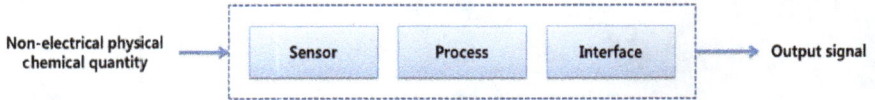

Fig. 3.1.   Sensor technology.

that you have on your smartphone has sensors and pressure sensors enable automatic opening of the doors in supermarkets).

Sensors are very common in everyday life.[2] A system that can track the state of the world at multiple levels as humans go about their day-to-day activities is of interest both for conceptual and practical reasons. The ability to recognize and reason the activities of a person, the resulting physical state of the world, and the likely emotional state of the actors is at the heart of computational models of human intelligence. The most ambitious sensor-based day-to-day state estimation systems to date have been human activity recognition systems.

Nowadays, the main structures of sensors can be miniaturized by micro/nanotechnologies, using a microelectromechanical system (MEMS) or a nanoelectromechanical system (NEMS). These technologies based on microfabrication technology can offer an intelligent and integrated sensor system with small dimensions, lower power consumption, and high reliability. In particular, the smart and complex sensors that use single-sensor modules are continually being developed using modern Si-based technology. Table 3.1 shows the classification of sensors by their fundamental working principles.

To understand the sensor's performance, some important characteristics of sensors must be considered for use in the real world. Generally, the characteristics of sensors are as listed below:

- *Accuracy*: Error between the result of a measurement and the true value being measured.
- *Resolution*: The minimum detectable signal fluctuation.
- *Sensitivity*: The ratio between the changes in the output signal to a small change in the input physical signal. Slope of the input–output fit line.

Table 3.1. Classifications of different sensors.

Mechanical quantities	Thermal quantities	Electromagnetic/ Optical quantities	Chemical quantities
Displacement			
Strain			
Rotation		Voltage	
Velocity		Current	
Acceleration	Temperature	Frequency phase	Moisture
Pressure		Visual/images	
Force/torque	Heat	Light	pH value
Twisting		Magnetic	
Weight			
Flow			

- *Repeatability/Precision*: The ability of the sensor to output the same value for the same input over a number of trials.
- *Dynamic Range* (*DR*): The ratio of maximum recordable input amplitude to minimum input amplitude, i.e., DR = 20 log (max. input amp./min. input amp.) in dB.
- *Linearity*: The deviation of the output from a best-fit straight line for a given range of the sensor.
- *Transfer Function* (*Frequency Response*): The relationship between physical input signal and electrical output signal, which may constitute a complete description of the sensor characteristics.
- *Bandwidth* (*BW*): The frequency range between the lower and upper cut-off frequencies, within which the sensor's transfer function is constant gain or linear. The BW of a sensor is the frequency range between these two frequencies.
- *Noise*: Random fluctuation in the value of input that causes random fluctuation in the output value. Therefore, all sensors produce some output noise in addition to the output signal. Many common noise sources produce a white noise distribution, which implies that the spectral noise density is the same at all frequencies. Johnson noise in a resistor is a good example of such a noise distribution.
- *Hysteresis*: Some sensors do not return to the same output value when the input stimulus is cycled up or down. The width of the

expected error in terms of the measured quantity is defined as the hysteresis.

2. Resistive Type Sensor

When electrons flow through a material, they experience some form of resistance, such as friction, that hinders their motion or free flow. Without such resistance, the electrons would accelerate since there is a force acting on them. The resistive force counterbalances the electric force; therefore, the drift velocity becomes constant. When the resistive force is high, then the current will decrease if the voltage difference that handles the motion remains the same. Then the resistance R will be

$$R \sim l/A, \tag{3.1}$$

where l is the path length and A is the cross-sectional area.[1]

Consider water flowing through a pipe. When the pipe is narrow, water flow is reduced. When the length is increased, the narrow pipe becomes longer. So, the flow is further reduced. In the same way, the voltage in a circuit is equivalent to the pressure of water flow and current is the flow rate. When the pressure (= voltage) difference increases, the flow rate (= current) will increase. When the electrical resistance increases, the current flow will decrease.

A resistive sensor device, based on Ohm's law, works depending upon the current flowing through it. It is required that all of the current applied at one node in the circuit flow to the same node. These two rules are called Kirchhoff's rules and are used to determine the currents and the voltages throughout a circuit. The first law is the Kirchhoff's current law, which determines the algebraic sum of the currents entering and the leaving any position in a circuit. Here are some examples of currents entering and leaving a point in a circuit in Fig. 3.2.

1. In Fig. 3.2(a), current I_a enters while I_b exits. Therefore the current is at this point is

$$I_a - I_b = 0. \tag{3.2}$$

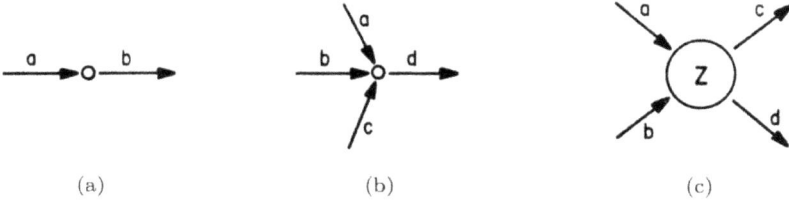

Fig. 3.2. Kirchhoff's first law.

2. Currents I_a, I_b and I_c enter while I_d exits in Fig. 3.2(b). Then the sum of the current is

$$I_a + I_b + I_c - I_d = 0. \tag{3.3}$$

3. Point Z or node Z has currents as shown in Fig. 3.2(c). Then the total current at Z becomes

$$I_a + I_b - I_c - I_d = 0. \tag{3.4}$$

The second rule is called as the Kirchhoff's voltage law, which expresses the algebraic sum of all voltages around a closed loop. When current flows through a resistor, a voltage is produced. The current enters the negative side of a component and exits on the positive to complete the electron flow. Here is an example of a circuit with current flow (Fig. 3.3).

For the basic circuit, we assume that the resistances (R_1 and R_2) of the material are constant, and circuit can be expressed using I and V. The linear relationship between V and I is called Ohm's law. Ohm's law can be expressed as:

$$V = I \times R. \tag{3.5}$$

Here, V is the voltage applied across the circuit in volts (V), I is the current flowing through the circuit in amperes (A), and R is the resistance of the circuit in ohms (Ω). Any material obeying Ohm's law is called an Ohmic material. According to Ohm's law, the linear relationship is confirmed between the voltage drop across a circuit and the current flowing through it. Therefore, resistance R is assumed to be constant, as a function of voltage and current.

2.1. *Basic Principles of a Resistive Sensor*

Now we turn our discussion to the resistive sensor. The following statements define the basic principles.

- The measured value directly or indirectly is changed according to the electrical resistance of a resistive material.
- Electrical resistance is a parameter to which voltage and current are related.
- Resistive sensing method is applied by taking advantage of changes in resistance compared to changes in other physical quantities. For example, in Fig. 3.4, we can assume a simple uniform conductor, $R = \rho \cdot L/A$ where ρ is the resistivity of material, L is the length,

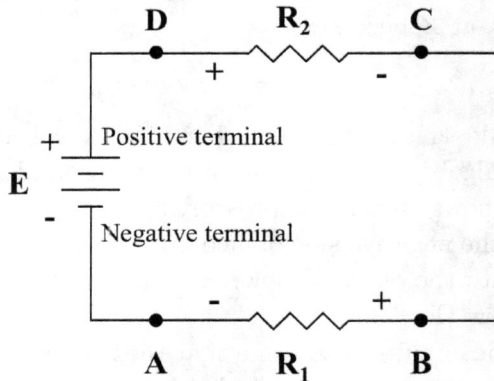

Fig. 3.3. An example of Kirchhoff's first law.

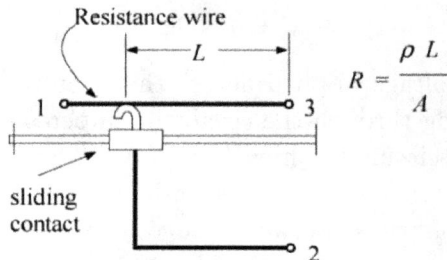

Fig. 3.4. A simple uniform conductor.

and A is the cross-sectional area through which the current flows. Then by moving the position of sliding contact, we can measure the resistance R as a function of length.

- Resistance can be changed either by a geometric (A, L) or material change in the resistive factor (ρ).
- Resistance is directly measured by an ohmmeter or through a signal conditioning circuit (e.g., a voltage divider).

2.2. *Wheatstone Bridge-Type Resistive Sensor*

A Wheatstone bridge circuit is a very general improvement on the voltage divider, as shown in Fig. 3.5. It consists simply of the same voltage divider, which is combined with a second divider composed of only fixed resistors. The additional divider point is added to ensure that the voltage is the same as the output of the sense voltage divider at some value of the sense resistance.

In Fig. 3.5, a circuit consisting of a power source and four resistors connected in a square. The resistors are connected to each other at nodes, which are labeled a through c. The circuit contains a potentiometer, labeled G, which detects the voltage difference between

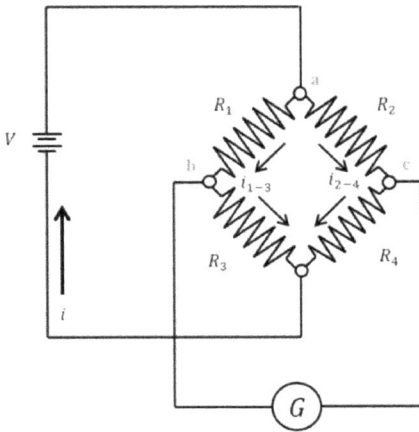

$$i = i_{1-3} + i_{2-4}$$

$$i_{1-3} = \frac{V}{R_1 + R_3}, \quad i_{2-4} = \frac{V}{R_2 + R_4}$$

$$V_b - V_a = \frac{V \cdot R_1}{R_1 + R_3}$$

$$V_c - V_a = \frac{V \cdot R_2}{R_2 + R_4}$$

$$G = V_c - V_b$$

$$\frac{G}{V} = \frac{R_2 \cdot R_3 - R_1 \cdot R_4}{(R_1 + R_3) \cdot (R_2 + R_4)}$$

Fig. 3.5. Wheatstone bridge circuit for a resistive sensor.

nodes c and b. The value from the potentiometer is displayed in the control room. If we consider each resistor separately, they have their own current (i_1, i_2, i_3, and i_4), resistance (R_1, R_2, R_3, and R_4), and voltage (V_1, V_2, V_3, and V_4) values, which are related to each other through Ohm's law. Resistors R_1 and R_3 are connected in series through node b. Therefore, the same current flows through R_1 and R_3,

$$i_{(1-3)} = i_1 = i_3, \tag{3.6}$$

and the value of $i_{(1-3)}$ can be determined from Ohm's law:

$$i_{(1-3)} = \frac{V}{(R_1 + R_3)}. \tag{3.7}$$

Similarly, resistors R_2 and R_4 are connected in series and the same current $i_{(2-4)}$ flows through these resistors. The current is given by

$$i_{(2-4)} = \frac{V}{(R_2 + R_4)}. \tag{3.8}$$

The change in voltage from node a to node b is given by

$$V_b - V_a = i_{(1-3)}R_1 = V \times \frac{R_1}{(R_1 + R_3)}. \tag{3.9}$$

Similarly, the voltage change from node a to node c is given by

$$V_c - V_a = i_{(2-4)}R_2 = V \times \frac{R_2}{(R_2 + R_4)}. \tag{3.10}$$

The potentiometer G measures the difference in voltage between nodes b and c.

$$G = V_c - V_b = (V_c - V_a) - (V_b - V_a), \tag{3.11}$$

$$G = V \times \left[\left\{ \frac{R_2}{(R_2 + R_4)} \right\} - \left\{ \frac{R_1}{(R_1 + R_3)} \right\} \right], \tag{3.12}$$

$$G/V = \frac{[(R_2 R_3) - (R_1 R_4)]}{[(R_1 + R_3)(R_2 + R_4)]}. \tag{3.13}$$

This final equation (3.13) explains how a Wheatstone bridge circuit can be used to eliminate temperature bias when using a strain

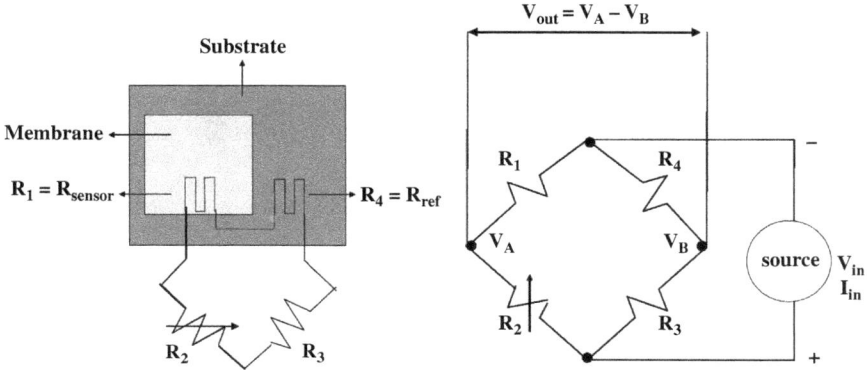

Fig. 3.6. Basic structure of piezo-resistive pressure sensor.[3]

gauge to determine forces on a wind tunnel model. Two strain gauges are connected to the model, and the output from the gauges are put into a Wheatstone bridge as R_1 and R_2. Equal "ballast" resistors are placed as R_3 and R_4. If the gauge is subjected to an increase in temperature, the resistance in both R_1 and R_2 increases by the same amount. But because the potentiometer measures the difference in resistance between R_1 and R_2, the reading stays the same.

Now, we introduce some representative types of the resistive sensor. The first one is a piezo-resistive pressure sensor (Fig. 3.6). In order to enhance the signal-to-noise (S/N) ratio of piezo-resistive pressure sensor, design optimization is commonly performed by considering different noise components.

Piezo-resistor integrated in the membrane is another type of sensor.[3,4] When pressure is applied to the membrane, it deflects the membrane. The membrane includes a resistive element that can respond to the applied pressure by changing resistance due to a structural deflection of the membrane. Then resistance change of resistive elements can be measured with Wheatstone bridges. The detailed concept is depicted in Fig. 3.7.

Another type is a piezo-resistive microsensor that makes use of oil damping. It consists of cantilever beams, a seismic mass, and oil.

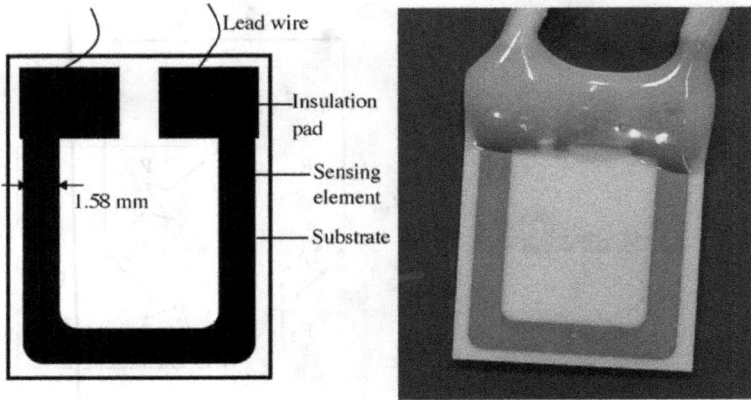

Fig. 3.7.   Thin film sensor.[4]

In the sensing structure, oil dampens the resonance of the suspended mass.

The third one is an electrical resistive sensor which can measure the *in situ* moisture content in municipal solid waste. The application of this technique can be used in bioreactor landfills where high moisture contents are expected and desired.

Recently, a novel resistive-type disposable humidity sensor on paper was reported. The main benefit of the printed sensor is that it is very cheap and easy to fabricate by inkjet printing using a conductive ink on paper substrate (Fig. 3.8). This disposable resistive-type moisture sensor can be used for checking the condition of goods during their transport or monitoring the living environmental conditions. Figure 3.8 shows the printed two chip-based semi-passive sensor.[5]

Also, a disposable humidity sensor on cellulose paper using single-walled carbon nanotubes functionalized with carboxylic acid demonstrated good repeatability and low hysteresis. The conductance shift of the nanotube network entangled on the microfibril cellulose is utilized for humidity sensing. Compared to the control sensor made on a glass substrate, the cellulose-mediated charge transport on the paper substrate enhances the sensitivity. This idea is a step toward for paper electronics in future for developing low-cost disposable sensor

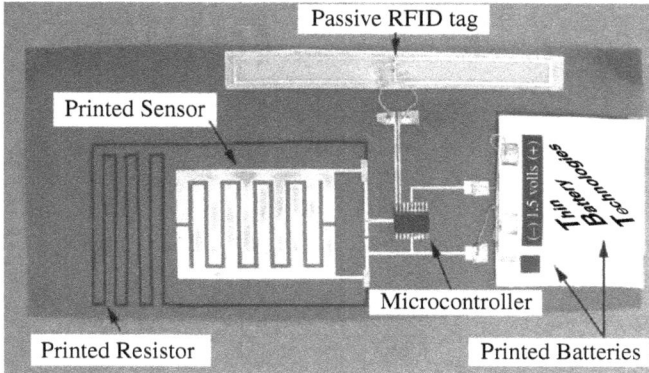

Fig. 3.8. Disposable printed sensor for the detection of moisture in packaging surveillance application.[5]

Fig. 3.9. Cellulose-based disposable humidity sensor.[6]

applications. The image in Fig. 3.9 shows the fabricated disposal sensor on paper.[6]

3. Capacitive Type Sensor

Capacitive sensing is a popular technology that has come to replace optical detection methods and mechanical designs for applications like proximity/gesture detection, material analysis, and liquid level sensing. The main advantage that capacitive sensing has over other detection approaches is that it can sense different kinds of materials (skin, plastic, metal, and liquid). Thus, it is contactless and wear-free and has the ability to sense up to a large distance with small sensor

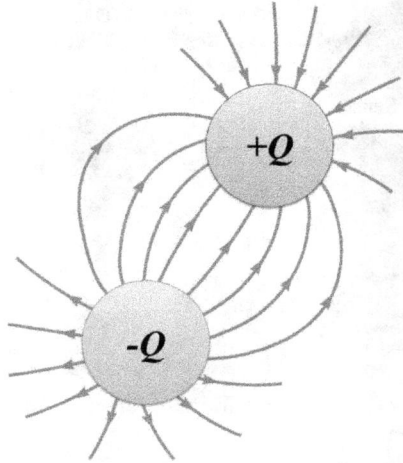

Fig. 3.10.   Basic configuration of capacitor.

sizes. The PCB sensor is a low-cost device, and it affords a low-power solution. This part describes the basic theory of capacitive sensor technologies and their use in various kinds of sensors in industrial applications.

3.1. *Basics of Capacitive Sensor*

Consider two conductors which carry charges of equal magnitude with opposite sign, as shown in Fig. 3.10. The electrical energy is stored in this arrangement and the capacity of charge is related to the medium between the two conductors. One plate is connected to positive charge $(+Q)$ and the other plate is connected to the negative charge $(-Q)$. Such a combination of two conductors is called a capacitor, as shown in Fig. 3.11. The capacitor is a device consisting of two conducting plates separated by a non-conducting substance called a dielectric (ε_r).[7,8] Air, mica, ceramic, fuel, or other suitable insulating materials are used as dielectrics.

Once a voltage is applied to the plates of a capacitor, the conducting plates will start to store electrical energy until the capacitor matches with the source voltage. Electric charge is added to an

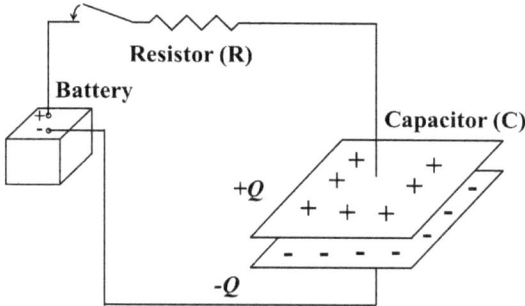

Fig. 3.11. Capacitor used in a circuit to store electrical charge.

isolated plate, and its potential is raised to V volts. The time to fully charge the capacitor is determined by time constant (τ),

$$\tau = RC, \tag{3.14}$$

where R is the resistor connected in the circuit and C is the capacitance to store the charges. Capacitance is measured in farad (F), which is defined as coulombs per volt. Thus, capacitance is an electrical property of the capacitors. If a given amount of charge is stored at a lower potential, V volts, this also means greater capacitance. Thus

$$C \propto Q \quad \text{and} \quad C \propto \frac{1}{V}, \tag{3.15}$$

$$\text{or} \quad C = \frac{Q}{V}, \tag{3.16}$$

where C is the capacitance in farad (F), Q is the magnitude of charge stored on each plate (coulomb), and V is the voltage applied to the plates (volts).

An electric field is formed due to the difference between the electric charges stored in the prospective plate surface, as shown in Fig. 3.12. Capacitance in this case depends upon the geometry of the conductors and not on an external source of charge or potential difference. In general, the capacitance value is determined by the

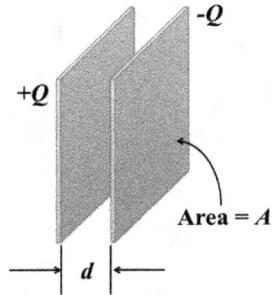

Fig. 3.12. A parallel-plate capacitor consists of two parallel conducting plates.

Table 3.2. Commonly used dielectric materials and their dielectric constants.[7,9]

Material	Dielectric constant	Material	Dielectric constant
Acetone	19.5	Mica	5.7–6.7
Air	1.0	Paper	1.6–2.6
Alcohol	25.8	Petroleum	2.0–2.2
Ammonia	15–25	Polystyrene	3.0
Carbon dioxide	1.0	Powdered milk	3.5–4.0
Chlorine liquid	2.0	Salt	6.1
Ethanol	24.0	Sugar	3.3
Gasoline	2.2	Transformer oil	2.2
Glycerin	47.0	Turpentine oil	2.2
Hard paper	4.5	Water	80.0

dielectric material, distance between the plates, and the area of each plate. Capacitance is given by

$$C = \varepsilon_r \frac{\varepsilon_0 A}{d}, \tag{3.17}$$

where C is the capacitance in farad (F), ε_r is the relative static permittivity (dielectric constant) of the material between the plates (dielectric constant of representative materials is summarized in Table 3.2), ε_0 is the permittivity of free space ($= 8.854 \times 10^{-12}$ F/m), A is the area of each plate, in square meters, and d is the distance (in meters) between the two plates.

The capacitance is strongly related to the electric field between the two metal plates of the capacitor. The electric field strength between the two plates decreases as the distance between the two conducting plates increases. Therefore, lower field strength or greater separation distance will lower the capacitance value. The conducting plates with larger surface area are able to store more electrical charge; therefore, a larger capacitance value is obtained with greater surface area.

3.2. *Capacitance in Parallel or Series Circuits*

The net capacitance of two or more capacitors, connected next to each other, depends on their connection configurations,[10] as shown in Fig. 3.13. If two capacitors are connected in parallel, they both will have the same voltage across them; therefore, their net capacitance will be the sum of the two capacitances. In parallel, the capacitors are connected as shown in Fig. 3.13.

Let us consider the electrical circuit. All top plates have the same potential V since they are all connected to the same wire (see point a, which has the potential V). Now look at the bottom part. Because they are all connected to the same wire, bottom plates have the same potential $-V$ (point b also has same potential). Then, we consider the net capacitance of a parallel combination of capacitors by charging as follows:

$$C_1 = \frac{Q_1}{V} \quad C_2 = \frac{Q_1}{V} \quad C_3 = \frac{Q_1}{V}. \tag{3.18}$$

Then total charge Q will be by summation of each charge:

$$Q_{\text{total}} = Q_1 + Q_2 + Q_3. \tag{3.19}$$

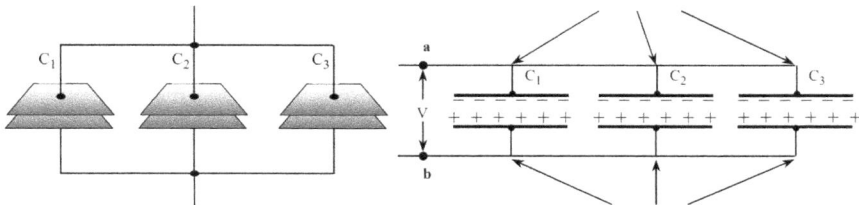

Fig. 3.13. Parallel connections of capacitors.

Total charge on a single capacitor C_{total} is equivalent to the combination of charge on the three capacitors C_1, C_2, and C_3.

Therefore, the total charge will be

$$C_{total}V = C_1V + C_2V + C_3V, \tag{3.20}$$

where C_{total} is the total capacitance of the capacitors connected in parallel. Finally, total charge is given by

$$C_{total} = C_1 + C_2 + C_3. \tag{3.21}$$

Figure 3.14 shows the capacitors connected in series. Two or more capacitors are connected in series; the voltage across the two terminals may be different for each capacitor although the electric charge will be the same on all of them. The equivalent capacitance of capacitors connected in series is calculated as follows. On each of the capacitors, C_1, C_2, and C_3, the left-hand plate induces an equal and opposite charge Q on the right-hand plate. Thus

$$Q_1 = Q_2 = Q_3 = Q_4. \tag{3.22}$$

The potentials across C_1, C_2, and C_3 are V_1, V_2, and V_3, respectively.

Between points a and b, the potential difference across the combination of the three capacitors becomes

$$V = V_1 + V_2 + V_3. \tag{3.23}$$

It can be rewritten as

$$\frac{Q}{C} = \frac{Q_1}{C_1} + \frac{Q_1}{C_2} + \frac{Q_1}{C_3}. \tag{3.24}$$

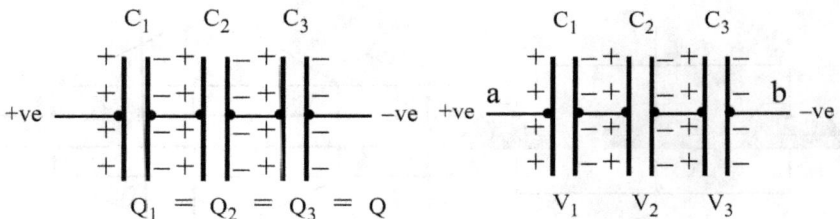

Fig. 3.14.   Capacitance in series.

Since $Q_1 = Q_2 = Q_3 = Q_4$, the total capacitance C will be

$$\frac{1}{C} = \frac{1}{C_1} + \frac{1}{C_2} + \frac{1}{C_3}.$$ (3.25)

3.3. *Capacitive Sensor*

Capacitive sensing is a capacitive coupling-based technology that takes the capacitance produced by the human body or other material as the input. It allows a more reliable solution for applications to measure liquid levels, material composition, mechanical buttons, and human-to-machine interfaces. A basic capacitive sensor is any metal or conductor used to detect difference between air and a conducting material or any material having a dielectric constant. Figure 3.15 shows three implementations for capacitive sensing: first is for proximity/gesture recognition, second for liquid level sensing, and third for material analysis.

These capacitive sensors have a wide variety of uses. Some categories of capacitive sensors are given below.

Flow measurement: A flow meter can convert a flow to pressure or displacement, using an orifice for volume flow or Coriolis effect force for mass flow. Capacitive sensors can then measure the displacement.

Pressure: A diaphragm with stable deflection properties can measure pressure with a spacing-sensitive detector.

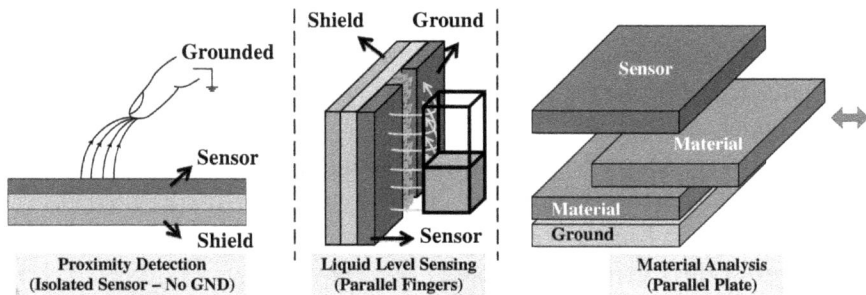

Fig. 3.15. Basic implements for capacitive sensing. (Adapted from Texas Instruments, Application Report SNOA927.)

Liquid level: Capacitive liquid level detectors sense the liquid level in a reservoir by measuring changes in capacitance between conducting plates, which are immersed in the liquid, or applied to the outside of a non-conducting tank.

Spacing: If a metal object is near a capacitor electrode, the mutual capacitance is a very sensitive measure of spacing.

Scanned multiplate sensor: The single-plate spacing measurement can be extended to contour measurement by using many plates, which are placed separately. Both conductive and dielectric surfaces can be measured.

Thickness measurement: Two plates in contact with an insulator will measure the insulator thickness if its dielectric constant is known or dielectric constant if thickness is known.

Ice detector: Icing on an airplane wing can be detected using insulated metal strips placed on the wing's leading edges.

Shaft angle or linear position: Capacitive sensors can measure angle or position with a multiplate scheme giving high accuracy and digital output, or with an analog output giving less absolute accuracy but faster response and simpler circuitry.

Lamp dimmer switch: The common metal-plate soft-touch lamp dimmer uses 60 Hz excitation and senses the capacitance to a human body.

Key-switch: Capacitive key-switches use the shielding effect of a nearby finger or a moving conductive plunger to interrupt the coupling between two small plates.

Limit switch: Limit switches can detect the proximity of a metal machine component through an increase in capacitance, or the proximity of a plastic component by virtue of its increased dielectric constant over air.

$X-Y$ tablet: Capacitive graphic input tablets of different sizes can replace the computer mouse as an $x-y$ coordinate input device. Finger-touch-sensitive, z-axis-sensitive and stylus-activated devices are available.

Accelerometers: Analog devices have introduced integrated accelerometer ICs with a sensitivity of 1.5 g. With this sensitivity, the device can be used as a tilt-meter.

Additionally, capacitive sensor also can be used for pressure, angular speed, and micro-sensing applications. Capacitive membrane pressure sensor can be used for the measurement of change in membrane capacitance, which assists in pressure monitoring, while a capacitive angular speed sensor has a fork arrangement and acts as a resonator where the resonator starts to oscillate when magnetic field and alternating current are applied (Lorentz force). The amplitude of the swing angle is detected by the capacitance change between movable and fixed electrodes. For sensing micro-sized capacitance, a capacitive cantilever micro-sensor, which consists of cantilevers acting as one electrode, an electrode strip, and a contact strip, is used. Saw-tooth voltage is applied to gradually increase the electrostatic force.

4. Impedance-Type Sensor

Impedance is the ratio of the voltage phasor to the current phasor. It is denoted as Z. The current–voltage relationship in AC circuits, in general, is given in the form[1,11]:

$$Z = V/I. \tag{3.26}$$

For a pure resistor, impedance is a resistor $(Z = R)$. Because the phase affects the impedance and the contributions of capacitors and inductors differ in phase from resistive components by 90°, a process like vector addition (phasors) is used to develop expressions for impedance. However, when impedance is applied to an AC circuit by including the frequency, real-world circuit elements become more complex and eventually exhibit resistive, capacitive, and inductive behaviors together, which is defined as impedance.

Generally, impedance consists of a real part (= resistance, R) and an imaginary part (= reactance, X). Reactance takes two forms: inductive (X_L) and capacitive (X_c). By definition,

$$X_L = 2\pi f L \quad \text{and} \quad X_c = 1/(2\pi f C), \tag{3.27}$$

where f is the frequency of interest, L is the inductance, and C is the capacitance.

Table 3.3. Common impedance measurement methods. (Adapted from Agilent Technologies.)

	Advantages	Disadvantages	Applicable frequency range
Bridge method	• High accuracy • Wide frequency coverage by using different types of bridges • Low-cost	• Needs to be manually balanced • Narrow frequency coverage with a single instrument	DC to 300 MHz
Resonant method	• Good Q accuracy up to high Q	• Needs to be tuned to resonance • Low impedance measurement accuracy	10 kHz to 70 MHz
$I-V$ method	• Grounded device measurement • Suitable to probe-type test needs	• Operating frequency range is limited by transformer used in probe	10 kHz to 100 MHz

Now, if $2\pi f$ is substituted by the angular frequency ω, then inductive and capacitive parts become $X_{\mathrm{L}} = \omega L$ and $X_{\mathrm{c}} = 1/\omega C$.

Using imaginary impedance values, impedance is now written as

$$Z = R + \mathrm{j}\omega L + 1/\mathrm{j}\omega C, \tag{3.28}$$

where j is the imaginary unit and ω is the angular frequency of the signal. Now, the impedance of the inductors is $\mathrm{j}\omega L$ while that of the capacitors is $1/\mathrm{j}\omega C$. In Table 3.3, we summarize common impedance-based measurement methods.

4.1. *Basics of Impedance Sensor*

The basic principle of impedance-type sensor is to measure the variations in resistance of the frequency response inductive or capacitive type sensors for displacement monitoring.[1] Generally, three different types of impedance sensors can be considered. (1) Capacitive type sensor is mainly used in a non-contact meter to monitor the precise measurement of a conductive material's position and also to measure the thickness of nonconductive materials. (2) Inductive-type sensor is also used for non-contact type displacement measurement using an eddy current detection. Therefore, it is not affected by any intermediate medium between the sensor and the target gap. (3) Capacitive and inductive type sensors can be used for relative non-contact positioning measurement, thickness measurement, and deflection and deformation of materials.

Recently, yttria-stabilized zirconia (YSZ) based on impedance metric sensor was reported to selectively detect propene (C_3H_6) under wet conditions. In these sensors, the sensing performance is very unique and much different from those of the potentiometric and ampere metric gas sensors reported so far. It has been confirmed that an impedance change (gas sensitivity) is attributed to the change in the resistance of electrode reaction at the oxide–YSZ interface. Figure 3.16 shows the YSZ-based impedance sensor and its schematic diagram.[12]

Impedance-type sensor also finds application in the field of medicine. It can be used to design non-contact vital signal sensors that can detect body signals such as respiration, heart beat, and so on. For example, a vital signal sensor works based on impedance variations and it consists of a resonator, an oscillator, a SAW filter, and a power detector. Figure 3.17 shows the basic concept of non-contact vital signal sensors using impedance measurement.[13] As a function of distance between human body — as a dielectric medium — and the resonator, the variation of the input impedance of a resonator leads to modification of the resonance frequency of the resonator, resulting in an oscillating frequency change. Using SAW filter, the sensitivity of the system can be improved. Using these devices, it is possible to monitor the respiration and heart rate within the maximum distance of 120 mm.

Monitoring glucose levels in realtime throughout the day for controlling blood sugar is necessary for patients with diabetes. Several companies are now engaged in developing impedance type sensors for

(a)

(b)

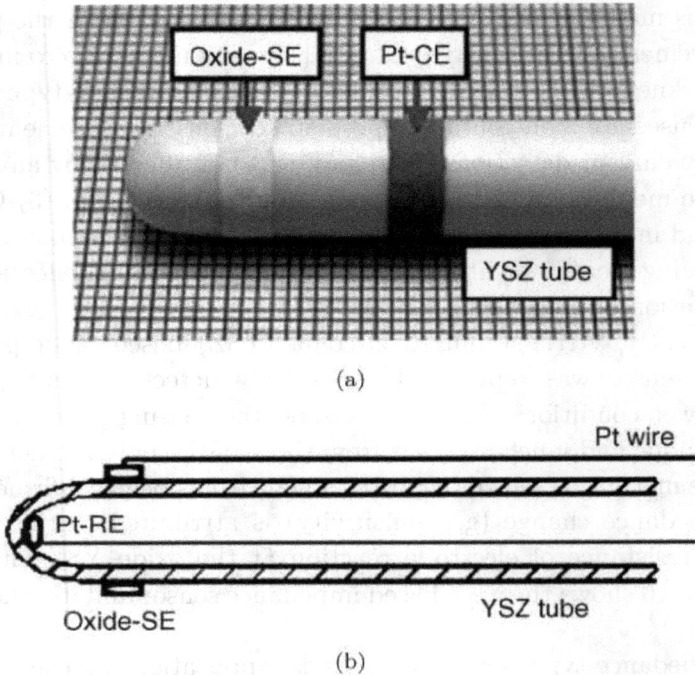

Fig. 3.16.   YSZ-based impedance metric sensor (a) and a cross-sectional view (b) of the tubular YSZ sensor attached with oxide sensing-electrode (SE).[12]

measuring glucose levels (Fig. 3.18). Using this idea, a non-invasive blood glucose meter called *Smart band* was designed to analyze glucose levels in blood.

This sensor detects the blood glucose level through the variation of impedance values that depend on the dielectric properties of blood. The change of glucose level in human body can trigger the dielectric characteristics of skin, which is detectable by impedance spectroscopic method. Using this concept, an impedance spectroscopic glucose meter called *glucoband* was developed. However, impedance spectroscopy is very sensitive to the measurement environments (e.g., temperature change, skin humidity, sweat, and micro flow in blood). Therefore, some correction methods are necessary to determine the correct value of glucose levels.

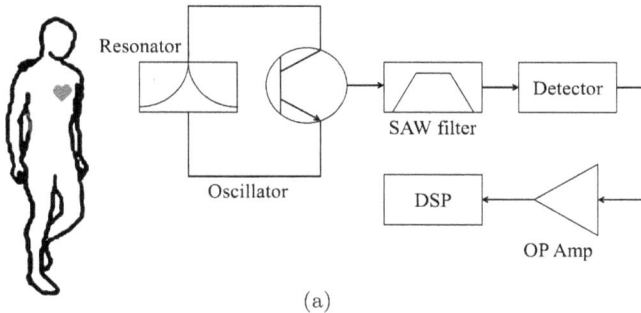

(a)

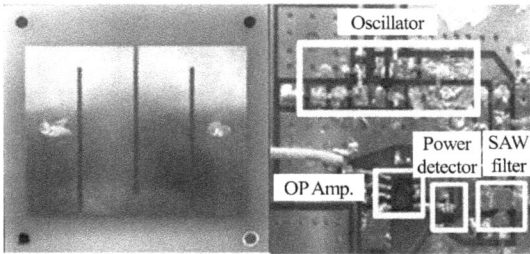

(b)

Fig. 3.17. (a) Concept of the proposed impedance medical sensor and (b) images of fabricated sensor system.[13]

Jawbone up series is also a smart band using the impedance parameter, as shown in Fig. 3.19. This smart band was created with the following components: (1) a precise accelerometer that tracks steps and activity; (2) two temperature sensors that collect ambient and skin temperature, and (3) a bioimpedance sensor that collects heart rate, respiration rate, and galvanic skin response. Accelerometer and temperature sensors are fairly common these days.

However, bioimpedance sensor is a novel device that can measure the resistance of a biotissue to a tiny amount of electric current, thus capturing a wide range of physiological signals. Traditionally, bioimpedance analysis is used in clinical settings to measure human body composition, such as body fat with respect to body's lean body mass.

For more complex sensor applications, impedance sensor is introduced into the multifunctional endoscopic system. This system is

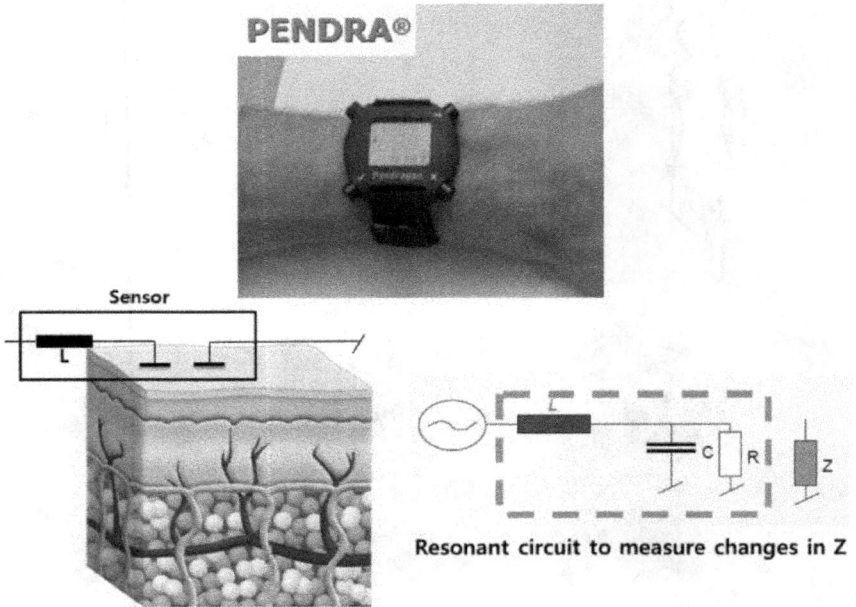

Fig. 3.18. The glucose meter and its equivalent circuit. (Adapted from PENDRA®.)

a multifunctional surgical endoscope system used to diagnose and treat intestinal diseases such as colon cancers. This "smart" endoscope system contains transparent bioelectronics, which provide impedance and pH-based sensing in combination with RF ablation therapy to facilitate the characterization and removal of colon cancers. Tumor/pH sensors, ablation electrodes, and viability sensors are calibrated and characterized *ex vivo* using both resected HT-29 (colon cancer cells) tissues and healthy tissues excised from the BALB/c (albino) nude mouse model. The tumor sensor is able to differentiate HT-29 tissues from normal tissues based on the difference in their impedances. The detailed structure of multifunctional endoscopic system is shown in Fig. 3.20.[14]

Recently wearable sensor technologies have been developed and used for personal health monitoring. Very recently, using on-site

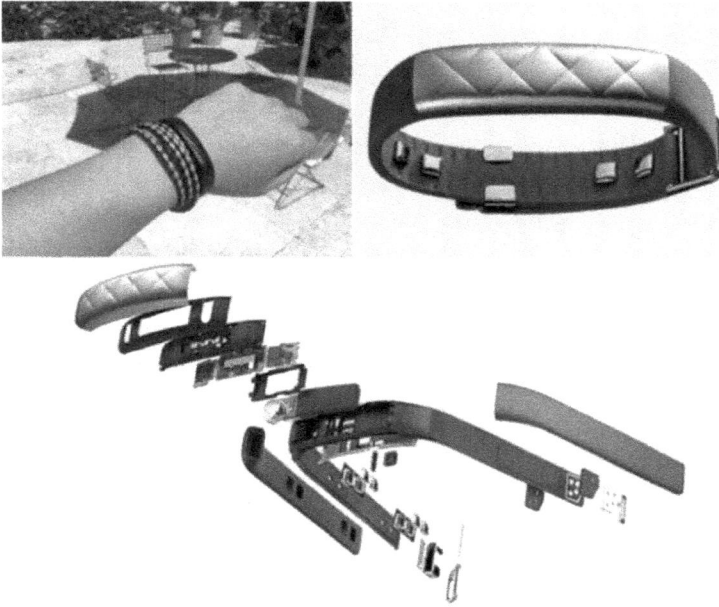

Fig. 3.19.   Smart band using bioimpedance. (Adapted from Jawbone up.)

signal processing circuitry and calibration method, a sweat-based non-invasive biosensor was developed as a fully integrated health monitoring system on a disposable substrate. A wearable flexible integrated sensing array (FISA) on a flexible printed circuit board (FPCB) for simultaneous and selective screening of a panel of biomarkers in sweat can be developed using the existing technological gap between signal transduction, conditioning (here, amplification and filtering), processing, and wireless transmission.[15] For the FISA system shown in Fig. 3.21, amperometric glucose and lactate sensor with current output are based on glucose oxidase and lactate oxidase immobilized within a permeable film of the linear polysaccharide chitosan. An Ag/AgCl electrode serves as a reference electrode and counter-electrode for both sensors.

Also, instead of expensive electronic sensors, a low-cost, low-power, and small size capacitive sensor to detect CO_2 was suggested

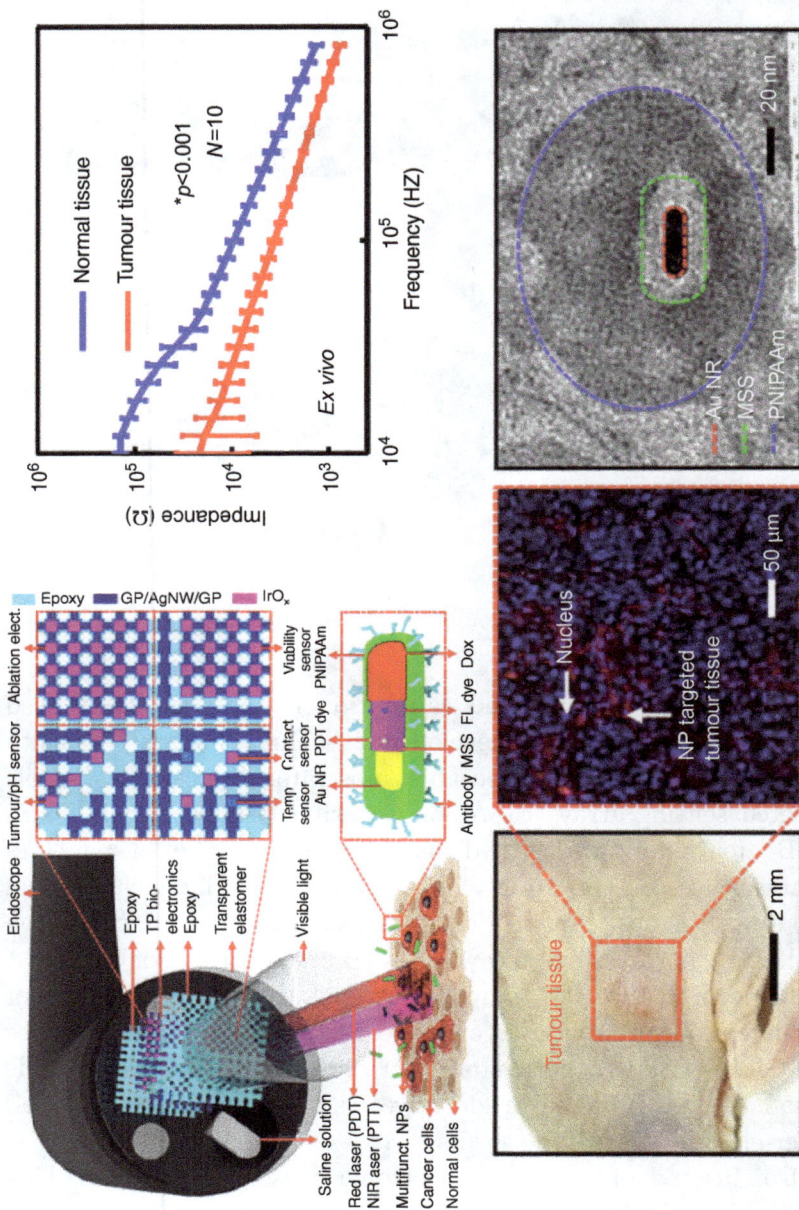

Fig. 3.20.   Multifunctional endoscopic system with impedance sensing.[14]

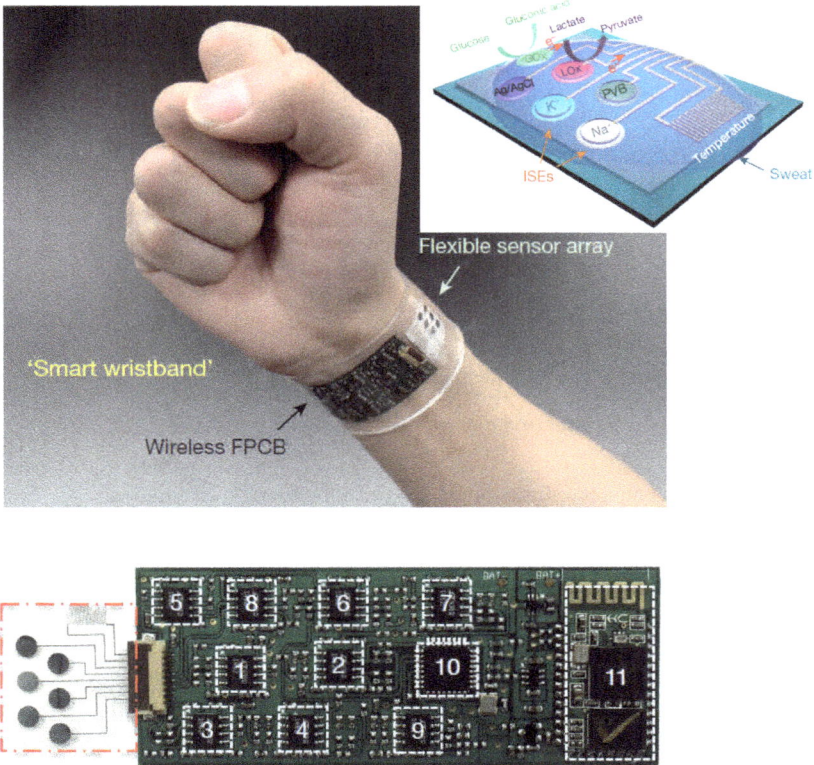

Fig. 3.21. Images of FISA for perspiration analysis (GO_x and LO_x, glucose) by integrating the multiplexed sweat sensor array and wireless FPCB.[15]

using dielectric nanocomposite comprising P-type multiwalled carbon nanotube and low molecular weight (Mw) organic $(1R, 2R)$-(+)-1,2-diphenylethylenediamine (DPED) stabilized with a resin (Fig. 3.22).[16] This small-sized sensor with simple electrical operation can be utilized for controlling industrial processes and monitoring air quality, breath, and blood analysis for medical diagnosis, food quality monitoring, and as a portable gas detector for special personal protection in future.

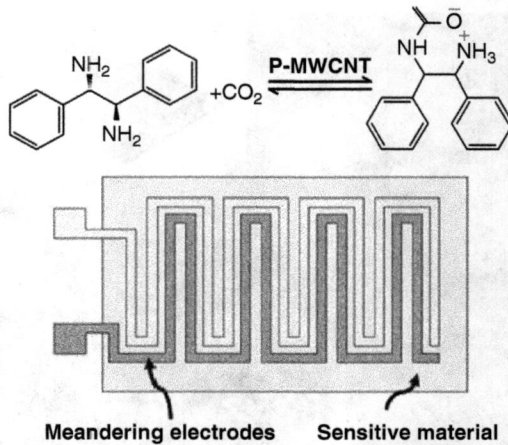

Meandering electrodes   Sensitive material

Fig. 3.22.   Schematic representation of reversible chemical reaction and the structure of capacitive sensor containing a pair of meandering electrodes.[16]

References

1. J. S. Wilson, *Sensor Technology Handbook*, Newnes, MA, USA (2004).
2. W. Pentney, A.-M. Popescu, S. Wang, H. Kautz and M. Philipose, Sensor-based understanding of daily life via large-scale use of common sense. In *Proc. 21th AAAI Conference for the Advanced Artificial Intelligence*, pp. 906–912, Boston, Massachusetts, USA (July 2006).
3. B. Bae, B. R. Flachsbart, K. Park and M. A. Shannon, Design optimization of a piezoresistive pressure sensor considering the output signal-to-noise ratio, *J. Micromech. Microeng.* **14**(12), pp. 1597–1607 (2004).
4. S. Li, S. Jung, K. W. Park, S. M. Lee and Y. G. Kim, Kinetic study on corrosion of steel in soil environments using electrical resistance sensor technique, *Mater. Chem. Phys.* **103**(1), pp. 9–13 (2007).
5. T. Unander and H.-E. Nilsson, Characterization of printed moisture sensors in packaging surveillance applications, *IEEE Sens. J.* **9**(8) 922–928 (2009).
6. J.-W. Han, B. Kim, J. Li and M. Meyyappan, Carbon nanotube based humidity sensor on cellulose paper, *J. Phys. Chem. C* **119**, 22094–22097 (2012).
7. R. A. Serway and J. W. Jewett, Capacitance and dielectrics, In *Physics for Scientists and Engineers*, 6th edn., Thomson, Scotland (2004).
8. E. Terzic, J. Terzic, R. Nagarajah and M. Alamgir, Capacitive and sensing technology, In *A Neural Network Approach to Fluid Quantity Measurement in Dynamic Environments*, Springer-Verlag, London (2012).
9. W. Benenson, H. Stoecker, W. J. Harris and H. Lutz, *Handbook of Physics*, Springer, New York (2002).

10. P. Scherz, *Practical Electronics for Inventors*, McGraw-Hill, New York (2000).

11. D. L. Mascarenas, M. D. Todd, G. Park and C. R. Farrar, Development of an impedance-based wireless sensor node for structural health monitoring, *Smart Mater. Struct.* **16**, 2137–2145 (2007).

12. M. Nakatou and N. Miura, Detection of propene by using new-type impedancemetric zirconia-based sensor attached with oxide sensing-electrode, *Sensors Actuators B: Chem.* **120**(1), 57–62 (2006).

13. K. Y. Kim, S. G. Kim, Y. Hong and J. G. Yook, Non-contact vital signal sensor based on impedance variation of resonator, *J. Korea Info. Commun. Soc.* **38**(9), 813–821 (2013).

14. H. Lee, Y. Lee, C. Song , H. R. Cho, R. Ghaffari, T. K. Choi, K. H. Kim, Y. B. Lee, D. Ling, H. Lee, S. J. Yu, S. H. Choi, T. Hyeon and D.-H. Kim, An endo-scope with integrated transparent bioelectronics and theranostic nanoparti-cles for colon cancer treatment, *Nat. Commun.* **6**, 10059 (2015).

15. W. Gao, S. Emaminejad, H. Y. Y. Nyein, S. Challa, K. Chen, A. Peck, H. M. Fahad, H. Ota, H. Shiraki, D. Kiriya, D.-H. Lien, G. A. Brooks, R. W. Davis and A. Javey, Fully integrated wearable sensor arrays for multiplexed *in situ* perspiration analysis, *Nature* **529**, 509 (2016).

16. M. Rahimabady, C. Y. Tan, S. Y. Tan, S. Chen, L. Zhang, Y. F. Chen, K. Yao, K. Zang, A. Humbert, D. Soccol and M. Bolt, Dielectric nanocomposite of diphenylethylenediamine and P-type multi-walled carbon nanotube for capacitive carbon dioxide sensors, *Sensors Actuators B: Chem.* **243**, 596–601 (2017).

Chapter 4

Chemical Sensors

Kyungbae Woo, Sangkyu Lee, Jae Eun Heo,
Seunghyeon Lee and Bong Sup Shim*

Department of Chemical Engineering, Inha University
100 Inha-Ro, Incheon 22212, Republic of Korea
*bshim@inha.ac.kr

Chemical sensors have been developed for many decades and are essential for our everyday modern life. They generally consist of a receptor and a transducer. Receptors recognize specific targets and then undergo some changes. This change causes a signal, and transducers convert this signal into data for quantitative and qualitative analysis. In this chapter, we introduce recent developments in receptor materials, paper-based sensors, biofriendly disposable sensors, pH sensors and colorimetric sensors. At the end of this chapter, we discuss chemical sensors of the future, in particular electronic nose and electronic tongue.

1. Introduction

Chemical sensors have been developed for many decades for a wide array of applications from medical diagnosis, to process control, to safety hygiene, and to environmental monitoring.[1] Because chemicals are essential elements for our everyday modern life, the use of chemical sensors is not just confined to fuels, food additives, and medicines, but is also expanded to materials, machinery, scientific research, and even war weaponry. Thus, a chemical sensor is now a fundamental tool for our safety from toxic chemicals, for preventing environmental pollution, and for process and quality control in manufacturing

77

plants. A chemical sensor basically transforms chemical information, such as molecular species, chemical activity, partial pressure and concentration, to analytically measurable signals, preferably electrochemical signals such as resistances, impedances, and capacitances. The most known examples of chemical sensors include monitoring toxic carbon monoxide (CO) gas released from incomplete combustion of heating fuels. This lethal, odorless CO gas is not detectable by humans, and therefore highly sensitive chemical sensors are necessary for their effective detection since even a low concentration in the air is enough to kill us silently. Smoke detectors, pollutant monitors and fire alarms are all equipped with these gas sensors. Other examples of sensors unrelated to gas sensing include glucose monitoring sensors as well as pregnancy test kits in which target analytes, glucose and human chorionic gonadotropin (hCG, the hormone indicating pregnancy), need to be selectively detected in blood and urine, respectively. These analytes require molecular-specific recognition and purifying channels for separating analytes from their complex mixtures. These sensors can be in the form of hand-held device for multiple use or as disposable kits for a quick one-time test. In addition, sensors can be used for the detection of chemical weapons such as sarin gas and anthrax-inducing bacteria, *Bacillus anthracis*, not just because they are lethal, but also because early detection for preventing an terrorist attacks requires extremely sensitive and selective sensing even at a single molecular level concentration.[2] Furthermore, various chemical sensors are essentially used in pollutant and smog monitors, weather forecasting systems, gas leakage alarms, environmental watches, etc. Recently, the chemical sensor technologies have been used for novel applications that have expanded to electronic noses and tongues for monitoring food poisoning to sensor arrays for analyzing complex chemical species with the development of high throughput functionalities made possible by advances in microelectromechanical systems (MEMS) and nanomaterials.[1,3,4] Continuous researches for molecular recognition, single-molecule detection, and wider sensor applications have been accelerated by the progress in nanomaterials and nanostructural composites, such as conducting polymers,[5] nanowires,[6,7] quantum dots, and, recently, graphene.[8] The various functional characteristics of newer materials significantly differ from the conventional sensor materials in terms of sensitivity, selectivity, and stability as well as applicability.

While the variety of chemical sensors is wide ranging, the definition of chemical sensors converges into the categories as suggested below.

1. Devices or instruments that determine the detectable presence, concentration, or quantity of a given analyte.[9]
2. Small-sized devices comprising a recognition element, a transduction element, and a signal processor capable of continuously and reversibly reporting a chemical concentration.[10] A small device, which, as the result of a chemical interaction or process between the analyte gas and the sensor device, transforms chemical or biochemical information of a quantitative or qualitative type into an analytically useful signal.[11]

The common notion of the suggested definitions of chemical sensors is a device that recognizes or detects matter (called "analyte"), transforming it into analyzable signals and thus providing the ability to monitor it in a perceivable manner.

The working principles of a chemical sensor is schematically presented in Fig. 4.1. A chemical sensor basically consists of two parts:

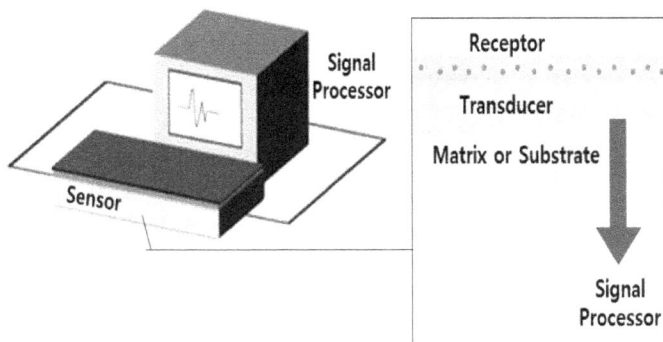

Fig. 4.1.   Schematic of the principle of a typical chemical sensor.

a receptor that recognizes a chemical target[12,13] and a transducer that causes changes in the signal on recognition of the target. Some researchers also include a processor that amplifies signals or filters noises from a transducer. Here is an example of how a sensor works. First, a receptor selectively recognizes the target analytes and undergoes chemical interactions with them. This process is called as "detection." For example, a receptor designed with tin oxide (SnO_2) is used for detecting methane, propane, and CO gases. The interactions that are used for detection include adsorption, physisorption, chemisorption, partitioning, acid–base neutralization, precipitation, ion exchange, absorption, addition, oxidation/reduction, chemical reaction, and, even decomposition. Next, a transducer transforms these interactions into electrical signals. The transduction involves complex electrical responsive actions of piezoelectric, potentiometric, photoelectric, or thermoelectric materials.[11,13] Magnetic field, optical light emission, electrochemical changes, mechanical responses, and thermal heat flow are also employed as signal transducing mechanisms in a chemical sensor. Then, the electrical signals are further transmitted to a user interface such as a display for our recognition.

Chemical sensors have many advantages compared to other conventional analytical equipment such as high-performance liquid chromatography (HPLC), mass spectroscopy (MS), or spectrophotometer.[10,13] They are relatively cheaper, smaller, and faster than these conventional analytical systems. Disposable, fully automatic, and high-throughput parallel sensors have already been commercialized so as to outperform the conventional equipment. Furthermore, with the advances in Internet of Things (IoT), the uses of sensors will grow multifold and the sensor that can be produced in a mass quantity will be preferred regardless of operating mechanisms. Thus, for mass production, disposability, biodegradability, and renewability as well as production costs become major issues for further development of chemical sensors. For sensor function itself, sensitivity, selectivity, and stability have always remained challenges that need to be fulfilled for basic sensor applications. Combined with development of information technology such as computers, data processing, and big data analytics, the sensor functions will be synergistically improved.[4,9,10]

In this chapter, we provide examples of disposable and flexible chemical sensors made with renewable materials such as biodegradable polymers and cellulose-based papers, discuss their manufacturing processes, performances, and enumerate their applications. We conclude the chapter with the suggestions for future research and development. We also review the paper-based diagnostics and electrical noses as emerging new applications of chemical sensors in the new wearable devices era.

2. Paper-based Sensors

In 1960s, Bazhenov and Fukada discovered that most materials made of wood possessed piezoelectricity, i.e., electric polarization under mechanical stress caused by internal rotation of carbon atoms adjacent to polar atomic groups.[14,15] Comparable to a quartz crystal, their unique piezoelectric properties originate from uniaxially oriented cellulose crystals in the wood fibers. These piezoelectric properties in cellulose papers were rediscovered in the 2000s and were suggested for applications as smart functional materials by J. H. Kim's research group.[16–18] Electro-active paper (EAPap), a cellulose-based actuating film, could be made by dissolving cellulose sources, casting a film into aligned cellulose crystalline structures, and coating them with thin electrodes on both sides of the surfaces.[19,20] (Fig. 4.2).

2.1. *EAPap Sensors*

The piezoelectric papers could find a wide range of applications as acoustic speakers, energy harvesters, stepper motors, wireless devices, and haptic sensors either in the form of pure cellulose or its composites.[21–26]

The EAPap system is also used in chemical sensor applications. Sungryul Yun *et al.* fabricated chemical vapor sensors made of

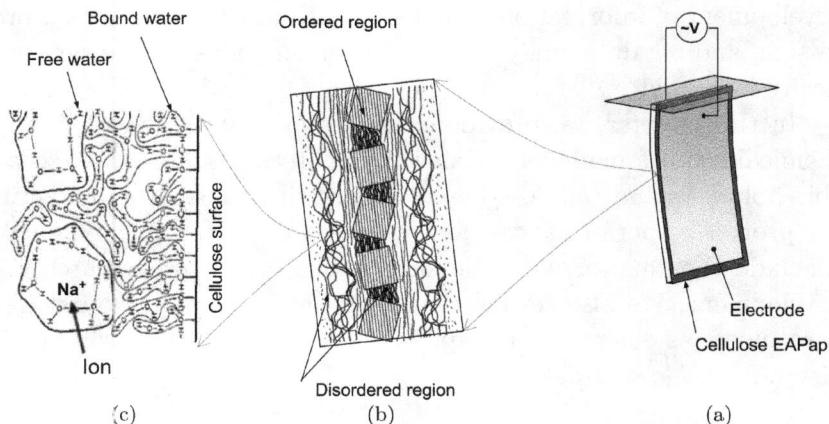

Fig. 4.2. Concept of electroactive paper(EAPap) actuator: (a) EAPap is consisted with the ordered crystalline regions and the disordered regions; (b) Gold electrodes are placed on both sides of the EAPap; (c) Water molecules are bounded on the cellulose surface.[16] (Reprinted with permission from Ref. 16. Copyright 2006 American Chemical Society).

cellulose films covalently grafted with multiwalled carbon nanotubes (MWCNTs). By transforming carboxyl groups into imidazolide, MWCNTs could be effectively grafted onto the cellulose film, which collected polar vapors selectively. The MWCNT–cellulose film sensors showed quantitatively titrated responses on the vapor concentrations of 1-butanol, 1-propanol, ethanol, and methanol. The sensors showed reproducible and reliable signals from repeated vapor exposures every 200 s.[27]

Mahadeva *et al.* used a cellulose nanocomposite on which polypyrroles are coated on the surface by *in situ* polymerization. The cellulose composites are used as flexible humidity and temperature sensors whose performances are linear and reversible in quantifying the analytes and the degrees. For stable sensing, it was essential to coat cellulose strands with polypyrrole to make it a uniformly integrated composite.[28]

Mun *et al.* demonstrated the use of cellulose–TiO$_2$–CNT hybrid nanocomposites as a gas sensor. The sensor was manufactured

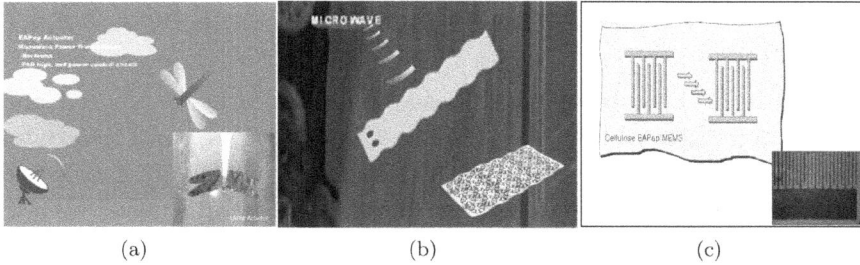

(a)   (b)   (c)

Fig. 4.3. Cellulose paper applications: (a) micro-flying object; (b) micro-insect robots; (c) biodegradable MEMS.[16] (Reprinted with permission from Ref. 16. Copyright 2006 American Chemical Society).

by blending celluloses with TiO_2–CNT, which are prepared by hydrothermal technique, and then patterning them. This sensor showed faster response and recovery of ammonia (NH_3) gas than other references, as the concentration of ammonia gas is varied from 50 to 500 ppm. The key features of the sensor include flexibility, cost, and disposability as well as good sensitivity and repeatability.[29] The applications of the cellulose-based papers are not only confined to the chemical sensors discussed earlier (Fig. 4.3), but also expanded to the wider sensors that can be used to measure wireless mechanical strength, pH, and vibration, and even as biosensors.[30–34]

For example, Jang *et al.* suggested the probability of using EAPap as a mechanical strain sensor for structural health monitoring. They reported that the resonance frequency of the sensor shifts due to tensile strain. They concluded that this sensor is simple, cost-effective, bioimplantable, and can operate in harsh environmental conditions.[30]

Kim *et al.* fabricated a vibration sensor using EAPap. The sensor was manufactured by attaching EAPap sensor at the bottom of the cantilevered beam. This sensor showed clear output voltage, and its impulsive response accurately reflected the dynamic characteristics of the beam, especially the twisting mode of the beam.[33]

2.2. *Specialized Paper-based Sensors: Microfluid Sensor*

Currently, two types of paper-based sensors are actively used and researched. These are lateral flow assays (LFAs) and microfluidic paper-based analytical devices (μPADs), which are fabricated with microscale fluid channels on a paper substrate using MEMS technology.[35]

2.2.1. *Lateral Flow Assays (LFAs)*

A typical LFA consists of a sample pad, a conjugate pad, an absorbent pad, and a backing pad on a cellulose-based paper membrane. Nitrocellulose is preferred because it is hydrophobic, which is required for guiding flows along the control lines. After the sample is injected in drops, analytes are pulled by capillary force into the absorbent pad, and then sensing molecules such as antibodies are used to complete the assay. In the LFAs, two design techniques are mainly used: sandwich and competitive. In a sandwich type, conjugated particles form complexes with analytes. The complexes are carried away by the fluid and get bound to the capturing molecules, which favor an interaction with analytes in a test line. The remaining conjugated particles also form a control line with other capturing molecules after passing through a test line. In contrast, conjugated particles and capturing molecules are pre-deposited on the test line and control line, so there is no aggregation with analytes. They are bound on the test line. In general, sandwich-type LFAs are used for analytes with multiple antigen epitopes, while competitive type is designed to detect analytes with a single antigen epitope.[35–37]

LFA-based biosensors (Bio-AMD) are the representative examples of paper-based diagnostic sensors (Fig. 4.4).

2.2.2. *Microfluidic Paper-based Analytical Devices (μPADs)*

μPADs are MEMS technology-based sensors in which a low-cost, disposable, and flexible paper is employed as the substrate. MEMS device offers one of the platforms for detection and analysis. From

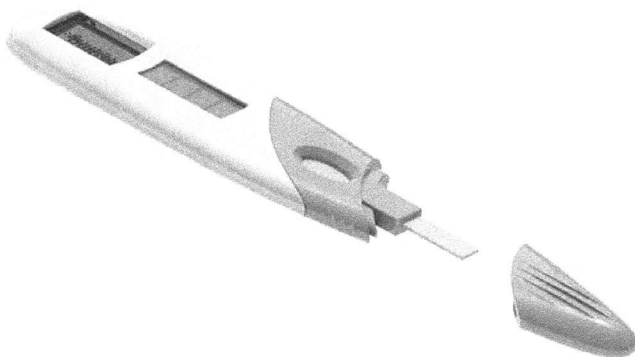

Fig. 4.4. LFA-based biosensor by Bio-AMD.[38]

glass or silicon to conventional polymers such as polydimethyl-
siloxane (PDMS) or SU-8, multiple classes of materials are used
as substrates. The MEMS devices are fabricated as 2D and 3D
structures and employed together with paper-based substrates. The
significant part of μPAD fabrication is a formation of micropat-
terned hydrophobic channels or walls in order to guide the samples on a hydrophilic paper substrate. A number of methods have
been developed.[39] Martinez *et al.* fabricated the first μPAD with
millimeter-scale patterned polymer channels on the paper by photolithography. The suggested that μPAD is one of the most promising candidates for POC determination of samples because it is inexpensive, portable, and simple. Furthermore, the sensor only requires
sample volumes as little as 5 μL and the performing time is as short as
10 min.[40,41]

A photolithography process is a sophisticated technique for
μPADs. Thus, it has disadvantages such as expensive instruments,
dangerous organic solvents, and complexity of the process. For over-
coming these, Lu *et al.* used wax printing techniques for manufacturing μPADs. They fabricated MEMS sensors by painting wax
on filter papers using a wax pen, inkjet printer, and wax printer.
Heat treatment (150°C) for 5–10 min was done to pattern hydrophobic channels, and they were able to assay horse radish peroxidase,

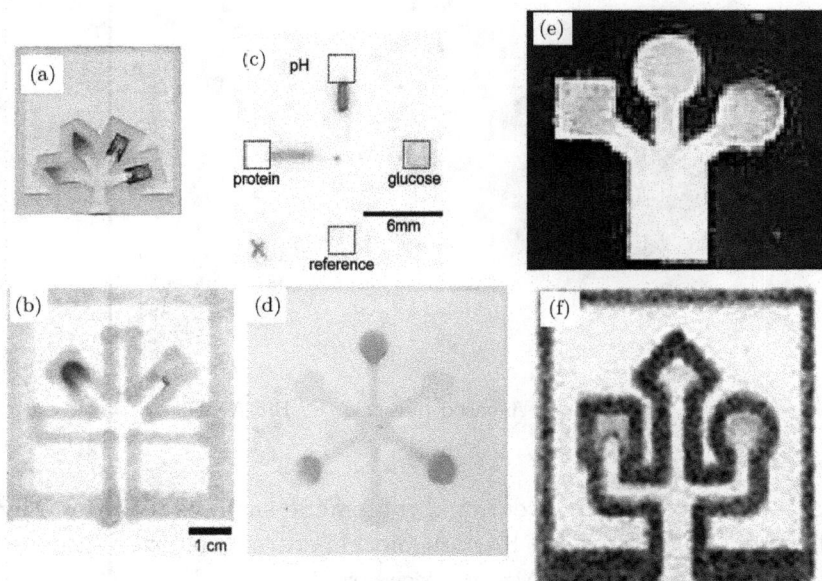

Fig. 4.5. μPADs fabricated by (a) photolithography,[41] (Reprinted with permission from Ref. 41. Copyright 2010 American Chemical Society); (b) plotting detecting,[44] (Reprinted with permission from Ref. 44. Copyright 2008 American Chemical Society); (c) inkjet etching,[45] (Reprinted with permission from Ref. 45. Copyright 2008 American Chemical Society); (d) plasma etching,[46] (Reprinted with permission from Ref. 46. Copyright 2008 American Chemical Society); (e) cutting detecting,[47] (Reprinted with permission from Ref. 47. Copyright 2009 American Chemical Society); (f) wax printing[48] (Reprinted with permission from Ref. 48. Copyright 2009 American Chemical Society).

bovine serum albumin, and glucose successfully using this method.[42] Temsiri et al. also fabricated μPADs that have hydrophilic channels by employing wax dipping on an iron mold.[43]

In addition to wax patterning, there are also many alternative techniques such as inkjet etching, plasma treatment, screen printing, and laser treatment for the fabrication of μPADs. The resulting sensor patterns are shown in Fig. 4.5.[43,45,46,49−52] If any device is not available in developing countries, even a direct handwriting with a BioPen, for example, on a paper can be used to fabricate POC diagnostic sensors as suggested by Han et al.[53]

3. Biofriendly Disposable Sensors

Conventional chemical sensors employ ceramic semiconducting substrates, which have no flexibility and biocompatibility. As the applications of the sensors are expanding, such as for wearable device applications, flexibility and biocompatibility become crucial issues for designing sensors. Thus, sensors using flexible substrates have received increasing attention. Furthermore, biodegradable and renewable materials carry unique potential for sensor applications compared to conventional petroleum-based polymers. New sensor applications such as disposable or biocompatible devices are also actively being explored along with making use of these novel environmentally friendly materials.

Among many renewable, environmentally friendly materials, poly(lactic acid), chitosan, cellulose nanocrystals, and their fibers are used for fabricating biodegradable functional sensors, usually combined with conductive carbon nanotubes (CNTs) and graphene.

3.1. *PLA-based Sensors*

Kumar *et al.* examined PLA/CNT nanocomposite-based chemical sensor for detecting volatile organic compound (VOC) gases using a layer-by-layer spray (sLBL) method. This sensor showed a high selectivity for chloroform in a mixture of four organic gases containing chloroform, methanol, toluene, and water.[54] (Fig. 4.6). Utilizing the conductivity of CNT, chemo-resistive PLA nanocomposite sensors are designed to transform electrical signals from chemical sensing. Because these four gases exhibit differences in solubility, polarity, and molecular size, their influence on the PLA/CNT sensors varied significantly, particularly causing a change in solubility parameters of the analytes on the matrix and percolative conductive networks of the composites.

3.2. *Chitosan-based Sensors*

Yu *et al.* used chitosan–Fe_3O_4 nanocomposite sensors for quantifying the concentration of bisphenol A (BPA). They claim that the

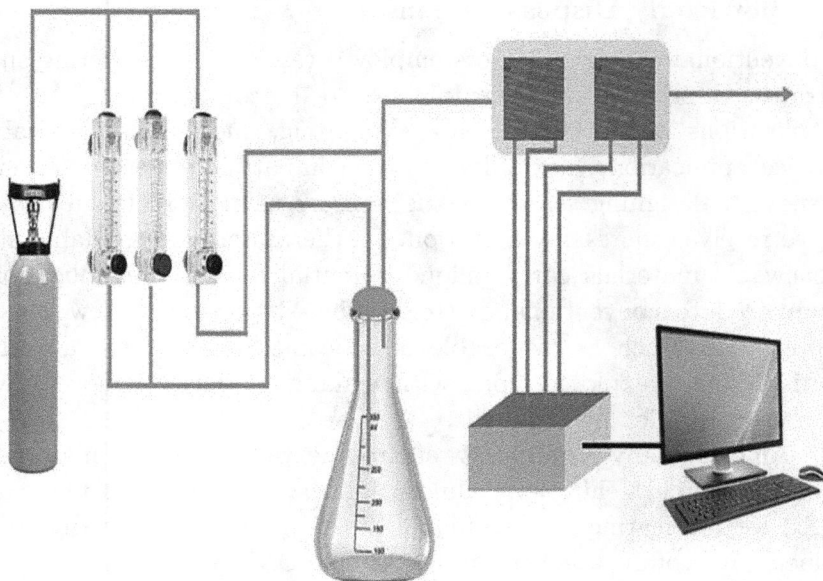

Fig. 4.6.   Schematics of chemo-resistive vapor sensor device.[54] (Redesigned the following concept of Ref. 54).

chitosan–Fe_3O_4 nanocomposites are unique in reducing oxidative overpotential, thus leading to an increase in electrochemical current responses, which improve the overall sensor functionality. These sensors could detect BPA at concentrations as low as 8.0×10^{-9} mol/dm^{-3} with S/N = 3. Furthermore, the sensors had a wide detection range compared to the other previous BPA sensors. In this example, chitosan is mainly used as matrix of Fe_3O_4 functional receptors. In contrast, there are other types of sensors using chitosan as a receptor. Mathew *et al.* used chitosan–gold nanocomposite as a sensor for detecting lead ions (Pb^{2+}) with a long-term stability of more than six months. The sensor showed high selectivity for lead ions in a mixed solution of various heavy metal ions such as cadmium, zinc, and lead. Furthermore, lead ions with concentrations as low as 10 μM could be detected by this sensor, which is sufficient for monitoring environmental samples with high reproducibility.[55] Nasution *et al.* also fabricated a chitosan film sensor for detecting acetone vapor, which is an important marker for diagnosis of diabetes from human breath. The

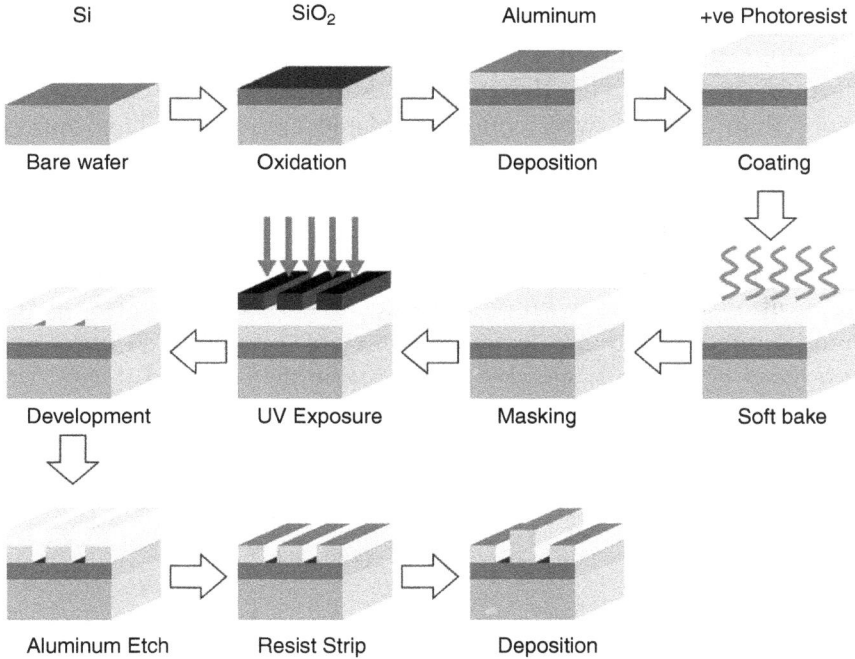

Fig. 4.7. Fabrication process of chitosan film sensor.[56] (Redesigned the following concept of Ref. 56).

chitosan gel was formed by electrochemical deposition on patterned Si wafers. This sensor could quantify acetone content in the range of 0.1–100 ppm at room temperature, while other conventional sensors could perform only at high temperatures, approximately 200°.

The sensors worked for more than 10 weeks without a drop in their robust performance levels.[56] (Fig. 4.7). Darder *et al.* designed potentiometric sensors for detecting anionic species by utilizing chitosan–clay nanocomposite as a reservoir of the active phase combined with conductive graphite particles. The sensor showed fast, durable, and reliable performance in detecting monovalent anions over multivalent anions for more than three months.[57]

3.3. *Cellulose-based Sensors*

Celluloses or their derivatives are also used as sensor materials. Liu *et al.* suggested that celluloses are useful on the microphotonic sensor platform. They coated 300-nm thick hydroxypropyl celluloses (H-HPC) on optical interferometer microfibers in order to prevent harmful bending. The H-HPC-coated microfibers were stable for more than 20 days with complex twisted structures. This optical microfiber acted as a refractive-index sensor with a sensitivity as high as 2600 nm/RIU.[58] Sadasivuni *et al.* prepared modified cellulose nanocrystal/graphene oxide film (m-CNC/GO) by layer-by-layer spraying for use as a photoelectric proximity sensor for touchscreen applications. Its sensor function was characterized by detecting the middle finger of humans from a distance of 0.2 mm. The film layer produced a uniformly charged electrostatic field around it, and the approaching finger induced voltage signal changes on the film layer. The sensitivity of m-CNC/GO film was five times higher in amplitude than that of a m-rGO film. The cellulose-based sensors are good candidates for new electronic applications because of their eco-friendly, transparent, non-metallic, and low-cost nature.[59]

4. pH Detecting Sensors

4.1. *Non-colorimetric pH Sensors*

Mahadeva *et al.* made cellulose and tin oxide hybrid composite as a disposable pH sensor. In this sensor, tin oxide is coated on a cellulose film via liquid phase deposition technique. The electrical conductivity is changed by the reaction between the tin oxide layer and the excess ions such as hydronium (H_3O^+) or hydroxide (OH^-). This sensor showed especially high detection ability in the pH range 4–8 and has potential for use in industrial applications.

4.2. *Colorimetric pH Sensors*

Devarayan *et al.* used sprayed natural color pigment (RC) from red cabbage on electrospun cellulose nanofibers (RC/Cs-ESNW) as a health monitoring pH sensor. Its RC color distinguished all the pH ranges quickly and smartly. This sensor allowed visible detection at pH values of 7–8 (the natural pH of human saliva) and 4–5 (the pH of saliva on consumption of alcoholic drinks). Thus, this RC/Cs-ESNW sensor can be used to monitor the level of alcohol intake.[60]

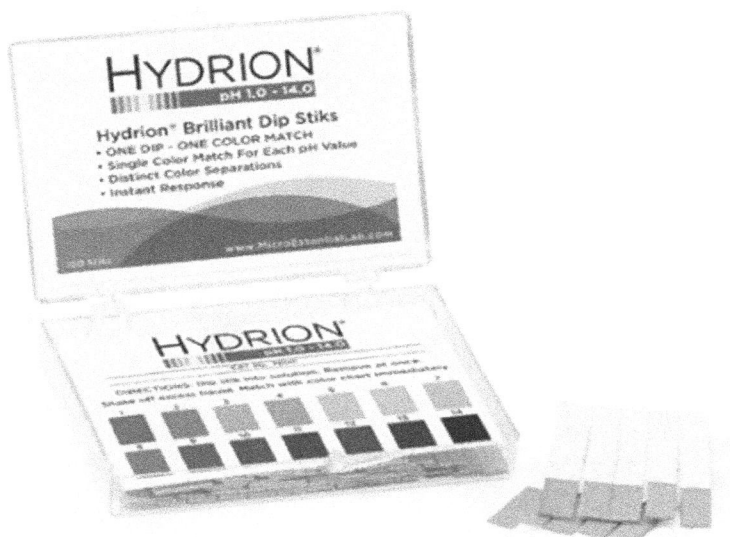

Fig. 4.8.   Commercial pH dipsticks by Hydrion®.

Dipsticks can be made simply by immersing test papers in acid–alkali indicator solutions. Starting from a glucose test paper by J. P. Comer in 1956, these dipstick papers have been mainly used for determining the pH of urine. pH paper is a typical example of these dipsticks.[61,62] (Fig. 4.8).

5. Colorimetric Sensor

Some reactions occur with color change. Colorimetric sensor, which works based on the color change phenomenon, enables qualification and quantification analysis by the naked eye. Also more accurate results can be obtained by using some optical detecting experiments, such as ultraviolet-visible spectroscopy (UV-vis spectroscopy) or emission spectroscopy.[63–66]

Cho *et al.* used several urea derivatives, which were synthesized by a reaction of 1,8-diaminonaphthalene and the corresponding isocyanates, to detect fluoride ions selectively (Fig. 4.9).

These compounds form a complex with F^-, and show a shift in the absorption peak in the presence of the fluoride ion. These

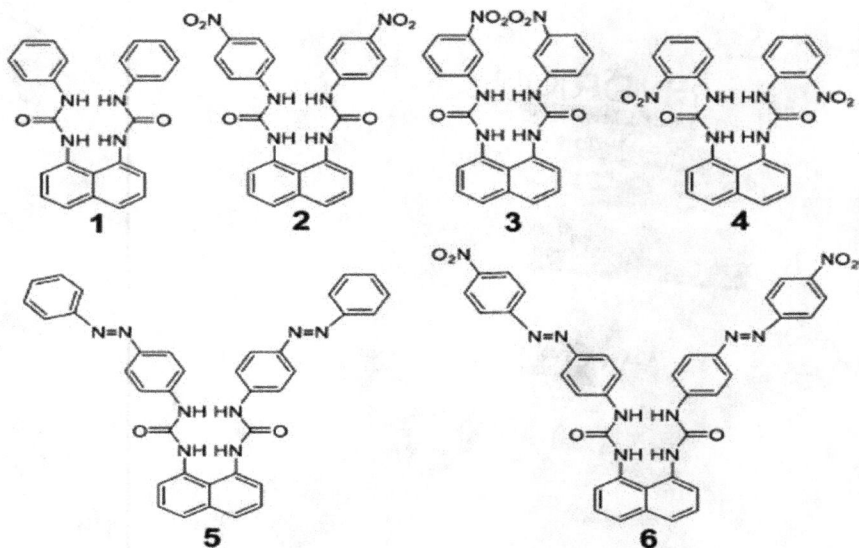

Fig. 4.9. Urea derivative naphthalene compounds.[67] (Reprinted with permission from Ref. 67. Copyright 2005 American Chemical Society).

ligands also show visible color changes in DMSO with the addition of tetrabutylammonium anions, and the result shows that these urea derivatives of naphthalene compounds could be used as a selective sensor for detecting fluoride.[67] (Fig. 4.10).

Ratnarathorn *et al.* examined the silver nanoparticle (AgNP) colorimetric sensor for the detection of Cu^{2+} ions by paper-based analytical devices. The homocysteine–dithiothreitol–AgNP solution was prepared using self-assembly of the aminothiol on the AgNP surface. Filter papers were modified with Hcy–DTT–AgNP solution. In the presence of Cu^{2+}, the intensity of plasmon resonance absorption peak at 404 nm decreased and a new red-shifted band at 502 nm was observed. When some metal ions at 200 times the Cu^{2+} concentration are added in the AgNP solution, an increase in the absorbance ratio was clearly observed but without any color change. Devices dried at ambient condition can be kept for 4 weeks at room temperature without loss of activity. As the World Health Organization (WHO) defines the maximum allowable levels of Cu^{2+}

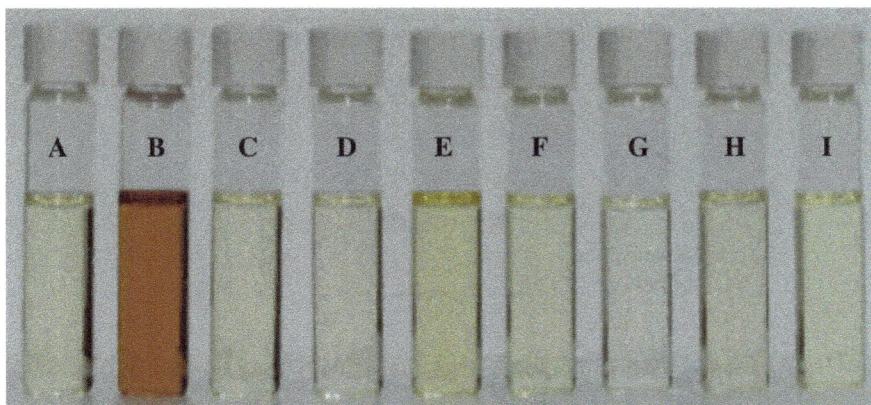

Fig. 4.10. Color changes of ligand 2 in DMSO with the addition of tetrabutylammonium anions. A=free receptor, B=fluoride, C=chloride, D=bromide, E=iodide, F=dihydrogen phosphate, G=hydrogen sulfate, H=acetate, I=benzoate.[67] (Reprinted with permission from Ref. 67. Copyright 2005 American Chemical Society).

in drinking water at 20.5 μM, this sensor having a sensing range of 7.8–62.8 μM Cu^{2+} should be useful for determining Cu^{2+} in drinking water.[68]

Colorimetric sensor can be easily employed to analyze many samples at the same time, so many researchers are engaged in developing a sensor array by using a colorimetric sensor. Dye–analyte interactions should be stronger than that of simple physical adsorption for colorimetric senor and usually three classes of chemoresponsive dyes are, used in a colorimetric sensor array: (1) Bronsted acidic or basic dyes (i.e., pH indicators), (2) Lewis acid/base dyes (i.e., metal ion containing dyes), and (3) dyes with large permanent dipoles (i.e., zwitterionic solvatochromic dyes).[69] (Fig. 4.11).

Janzen *et al.* developed a low-cost, sensitive colorimetric sensor array for detecting and identifying the VOCs.[69] One hundred commonly known VOCs were analyzed and detected by their distinct RGB color to differentiate each compound. Also, Zhang *et al.* designed a sensor array for detecting organics in water, which was used for testing common soft drinks, providing an example of

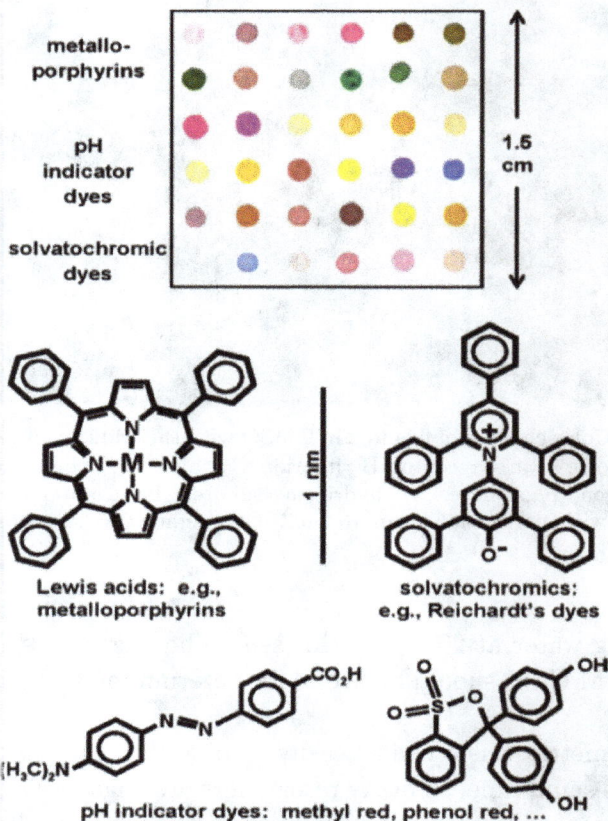

Fig. 4.11. Disposable colorimetric sensor array and its dye classes.[69] (Reprinted with permission from Ref. 69. Copyright 2006 American Chemical Society).

the application of colorimetric sensor in a real-world case. Each commercial soft drink is distinguished from others by a visible detection of its spot.[70] (Fig. 4.12).

6. Future Applications of Chemical Sensors: Electronic Nose and Electronic Tongue

While various chemical sensors have been developed for decades, their ultimate goal of application is to mimic and replace sensing

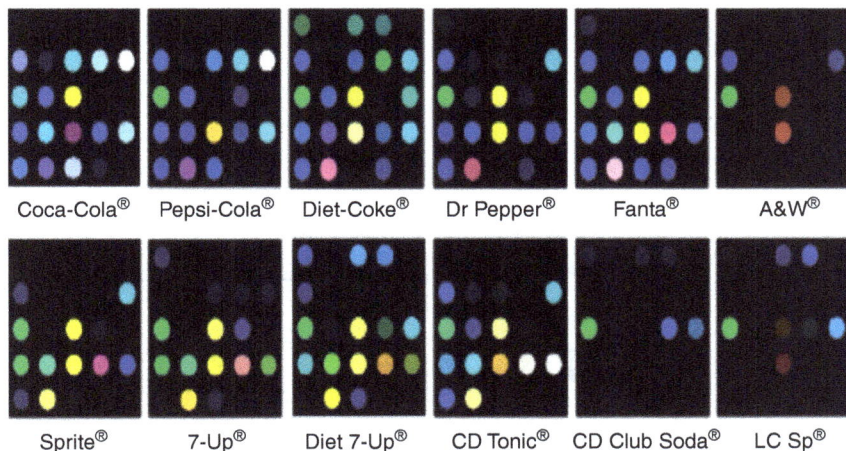

Fig. 4.12. Color change profiles with the acid-sensitive array for a common soft drinks.[70] (Reprinted with permission from Ref. 70. Copyright 2006 American Chemical Society).

organs such as animal/human noses and tongues, which can selectively detect a component from compounding sources.

6.1. *Electronic Nose* (*E-nose*)

E-nose is an assembly of chemical sensors playing the role of a mammalian nose. Following this defining concept, E-nose detects gases, analyzes their patterns, and recognizes them as a "fingerprint".[71,72] (Fig. 4.13).

Thus, E-nose commonly consists of a sensor module, a board, and an analyzing device such as a computer.[73,74] A sensor module is made with sensor arrays for detecting multiple components. The board collects information from these sensor modules. The analyzing device integrates the information and indicates or displays them as results.[75] (Fig. 4.14).

There are many applications of E-nose. A typical example is monitoring the fermentation process of black tea. There are some differences in the fermentation state of black tea, which varies with material species present in it. E-nose can be used to investigate

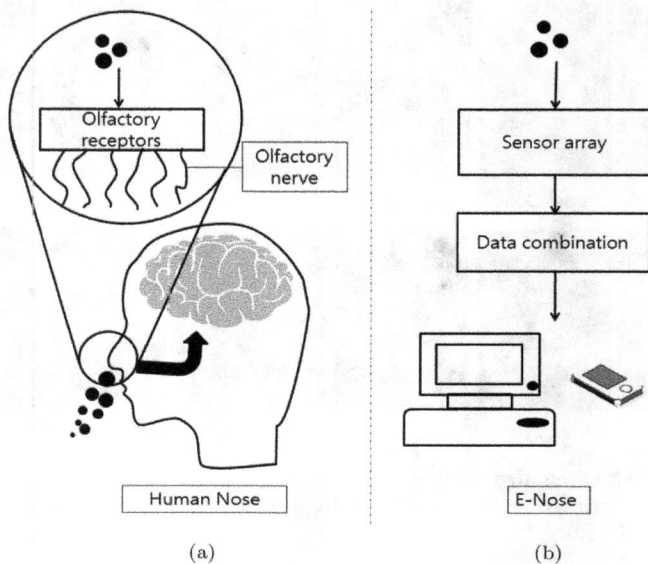

Fig. 4.13. Comparison between biological nose (a) and electronic nose (b).

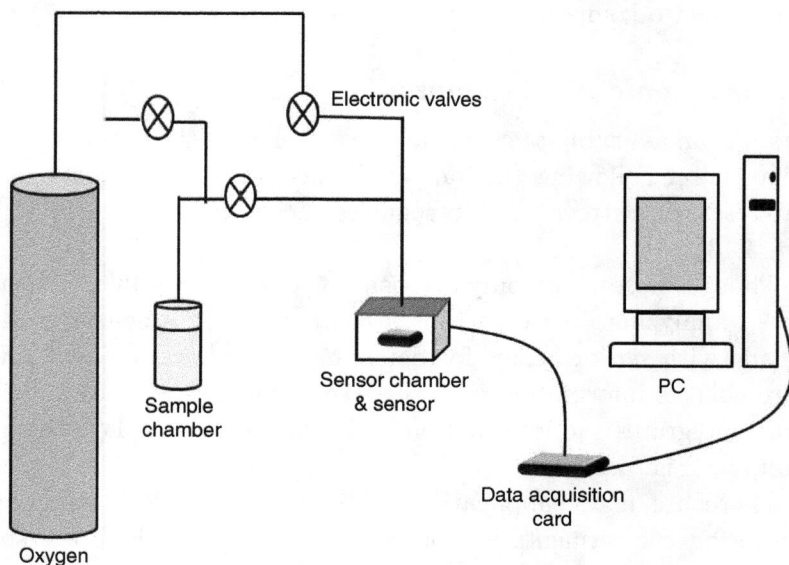

Fig. 4.14. Conventional E-nose system.[76] (Redesigned the following concept of Ref. 76).

various material species of black tea. Using the results, the optimum fermentation time for each tea can be determined. Like this example, E-noses have been used in the food and agricultural industry for monitoring food states, determining best time for harvesting fruits, and controlling the quality of saffron, an expensive spice.[76–79] E-noses are also used in medical care. It is known that specific volatile gases are emitted from an organ when it has diseases or problems in functioning normally. So, active research is ongoing to design E-noses for detecting these volatile gases, which human nose cannot detect, for early diagnosis or prevention of diseases.[80–83] In addition, E-nose can be used to monitor the contamination level of water by detecting toxic materials emitted from bacteria in a river or a lake.[84,85] They can also be employed for monitoring the air quality level indoors and outdoors.[86] and for the detection of odorless toxic gases using an alarm.[87]

With the advancement of technologies, complex sensor arrays are reduced to a tiny sensor chip module.[89] Complex analyzing equipment is replaced by a portable mobile device like personal digital assistant (PDA) or smartphones.[90] For example, a product named FOOD Sniffer (Fig. 4.15) can detect VOC gases, which can be

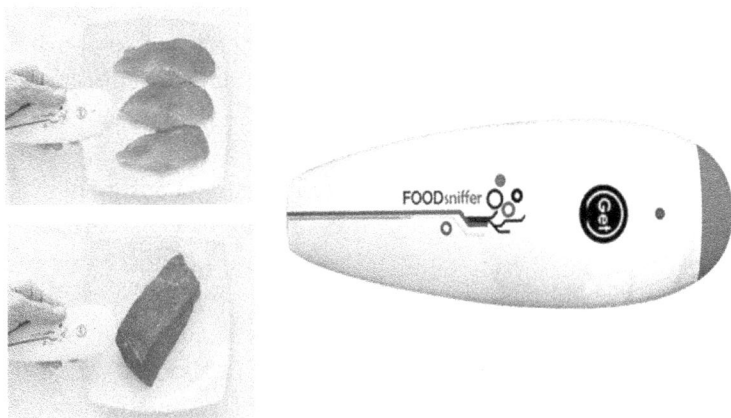

Fig. 4.15. Foodsniffer®; a recently commercialized portable E-nose device with a mobile phone.[88]

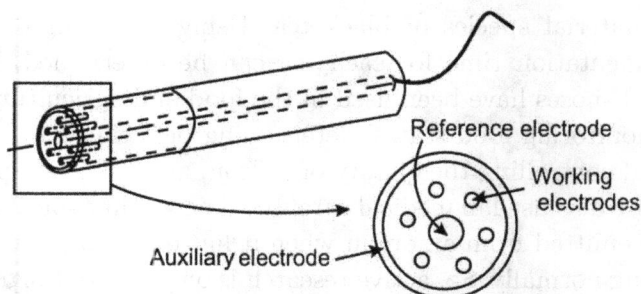

Fig. 4.16. Electrode arrangement in the voltammetric electronic tongue.[99] (Reprinted with permission from Ref. 99. Copyright 2000 Elsevier).

used to monitor the decay of foods by connecting the device to a smartphone.[91] Handheld E-noses are few now as commercialization of low cost, disposable miniaturized sensors is yet to be realized.

6.2. Electronic Tongue (E-tongue)

E-tongue is the analytical device that mimics the tongue in analyzing taste sensation. It is similar to E-nose in regard to analyzing patterns of the sample and making a "fingerprint." But E-tongue can also detect liquid phase substances unlike E-nose.[92-99] (Fig. 4.16).

Electrochemical sensors, including potentiometric, voltammetric, conductimetric, amperometric, and impedimetric sensors, are mainly used in the E-tongue sensor array. Among them, potentiometric sensors and voltammetric sensors are mainly used.

In the case of potentiometric sensors, a potential, which can determine the concentration of the component of interest in a solution, is measured without current flow. In the electrochemical cell, a free energy change occurs when the equilibrium condition is reached, and this free energy change gives rise to a potential.[92-95,98]

In voltammetry, a change in potential leads to a measurable change in current. A voltammetric sensor displays high sensitivity, simplicity, versatility, and robustness. However, all components in a measured solution have an effect on the measured current, and therefore selectivity is low in a voltammetric sensor.[96,97]

There are many applications of E-nose such as in food analysis and water pollution monitoring, among others.

Natale *et al.* used commercial chalcogenide glass sensors provided by Analytical Systems (St. Petersburg, Russia) and monitored samples of water from Neva River (a river flowing through the town St. Petersburg). They artificially added Cu, Cd, Fe, Cr, and Zn in ionic form and measured the total concentrations of Cu, Cd, Fe, Cr, Zn, Cl, SO_4, and H. Data was analyzed by multiple linear regression (MLR), partial least squares (PLS), nonlinear least squares (NLLS), and back-propagation-neural network (BP-NN) to improve the accuracy of the estimations.[98]

Legin *et al.* distinguished between different kinds of the same type of beverages (tea, coffee, juices, soft drinks, beer) and monitored the quality of fruit juices during storage and aging by using E-tongue.[93] Gil *et al.* designed an E-tongue formed by a 16-electrode array using a thick-film technology. Good correlation was found between the E-tongue and parameters such as total biogenic amines, pH, microbial analysis, and TVB-N at a relatively low cost and in less time.[95]

Since flavor is inferred not only through tongue for taste but also by nose for smell, some researchers have combined an E-nose with an E-tongue. Di Natale *et al.* used porphyrin-based quartz microbalance gas sensor array as an E-nose and potentiometric electrode sensors as an E-tongue. Urine and milk were analyzed as a representative case in a clinical analysis and in food analysis. Inspite of two very different fields, such as clinical and food anaylsis, improved classification performance was displayed by combining E-nose and E-tongue.[94]

Although novel technologies are being reported, there still exists a barrier for their commercial use. Thus, mass producible cheap sensor technology is required for wide applications of these sensors in our daily life.

7. Conclusion

As discussed in this chapter, there are new classes of promising low-cost sensors prepared making use of polymers, cellulose, and papers. Unlike conventional ceramic-based semiconducting chemical sensors,

sensors fabricated from these materials are flexible, environmentally friendly, and sustainable even in developing countries. In this chapter, we discussed recent research progress in chemical sensing receptors, paper-based sensors, biofriendly disposable sensors, pH sensors, and colorimetric sensors. As new technologies such as IoTs or electronic noses are emerging, the sensor technologies in general will further find new applications. Thus, chemical sensors designed using disposable and renewable materials will be ubiquitous in the near future.

References

1. J. R. Stetter and W. R. Penrose, Understanding chemical sensors and chemical sensor arrays (electronic noses): Past, present, and future, *Sensors Update* **10**(1), 189–229 (2002).
2. A. A. Tomchenko, G. P. Harmer and B. T. Marquis, Detection of chemical warfare agents using nanostructured metal oxide sensors, *Sensors Actuators B: Chem.* **108**(1–2), 41–55 (2005).
3. E. L. Hines, E. Llobet and J. W. Gardner, Electronic noses: A review of signal processing techniques, *IEE Proc. Circ. Dev. Syst.* **146**(6), 297–310 (1999).
4. X.-J. Huang and Y.-K. Choi, Chemical sensors based on nanostructured materials, *Sensors Actuators B: Chem.* **122**(2), 659–671 (2007).
5. J. Janata and M. Josowicz, Conducting polymers in electronic chemical sensors, *Nat. Mater.* **2**(1), 19–24 (2003).
6. H. Liu, J. Kameoka, D. A. Czaplewski and H. G. Craighead, Polymeric nanowire chemical sensor, *Nano Lett.* **4**(4), 671–675 (2004).
7. M. C. McAlpine, H. Ahmad, D. Wang and J. R. Heath, Highly ordered nanowire arrays on plastic substrates for ultrasensitive flexible chemical sensors, *Nat. Mater.* **6**(5), 379–384 (2007).
8. Y. Ju Yun, W. G. Hong, N.-J. Choi, B. Hoon Kim, Y. Jun and H.-K. Lee, Ultrasensitive and highly selective graphene-based single yarn for use in wearable gas sensor, *Sci. Rep.* **5**, 10904 (2015).
9. Committee on New Sensor Technologies: Materials and Applications, Commission on Engineering and Technical Systems, National Research Council *Expanding the Vision of Sensor Materials*, Washington, DC: National Academy Press, (1995).
10. O. Wolfbeis, Chemical sensors — Survey and trends, *Fresenius' J. Anal. Chem.* **337**(5), 522–527 (1990).
11. J. R. Stetter, W. R. Penrose and S. Yao, Sensors, chemical sensors, electrochemical sensors, and ECS, *J. Electrochem. Soc.* **150**(2), S11–S16 (2003).
12. K. J. Albert, N. S. Lewis, C. L. Schauer, G. A. Sotzing, S. E. Stitzel, T. P. Vaid and D. R. Walt, Cross-reactive chemical sensor arrays, *Chem. Rev.* **100**(7), 2595–2626 (2000).

13. C. A. Galán-Vidal, J. Muñoz, C. Domínguez and S. Alegret, Chemical sensors, biosensors and thick-film technology, *TrAC Trends Anal. Chem.* **14**(5), 225–231 (1995).

14. V. A. Bazhenov, *Piezoelectric Properties of Woods*, Consultants Bureau, New York (1961).

15. E. Fukada, History and recent progress in piezoelectric polymers, *IEEE Trans. Ultrason. Ferroelectr. Freq. Control* **47**(6), 1277–1290 (2000).

16. J. Kim, S. Yun and Z. Ounaies, Discovery of cellulose as a smart material, *Macromolecules* **39**(12), 4202–4206 (2006).

17. J. Kim, S. Yun and S. K. Lee, Cellulose smart material: Possibility and challenges, *J. Intelli. Mater. Syst. Struct.* **19**(3), 417–422 (2008).

18. J. Kim and Y. B. Seo, Electro-active paper actuators, *Smart Mater. Struct.* **11**(3), 355–360 (2002).

19. S. Yun, J. Kim and K.-S. Lee, Evaluation of cellulose electro-active paper made by tape casting and zone stretching methods, *Int. J. Prec. Eng. Manuf.* **11**(6), 987–990 (2010).

20. J. Kim, C. S. Song and S. R. Yun, Cellulose based electro-active papers: performance and environmental effects, *Smart Mater. Struct.* **15**(3), 719–723 (2006).

21. J. Kim, G. Y. Yun, J. H. Kim and J. Lee, Piezoelectric electro-active paper (EAPap) speaker, *J. Mech. Sci. Technol.* **25**(11), 2763–2768 (2011).

22. Z. Abas, H. S. Kim, L. Zhai, J. Kim and J. H. Kim, Electrode effects of a cellulose-based electro-active paper energy harvester, *Smart Mater. Struct.* **23**(7), 074003 (2014).

23. B. W. Kang and J. Kim, Design, fabrication, and evaluation of stepper motors based on the piezoelectric torsional actuator, *IEEE/Asme Trans. Mechatronics* **18**(6), 1850–1854 (2013).

24. S. Y. Yang, J. Kim and K. D. Song, Flexible patch rectennas for wireless actuation of cell lose electro-active paper actuator, *J. Electr. Eng. Technol.* **7**(6), 954–958 (2012).

25. S. Yun, G. Y. Yun, K. B. Kim, B. W. Kang, J. Kim and S. Y. Kim, Film-type haptic actuator made with cellulose acetate layers, *J. Intell. Mater. Syst. Struct.* **25**(11), 1289–1294 (2014).

26. G. Y. Yun, J. Kim, J. H. Kim and S. Y. Kim, Fabrication and testing of cellulose EAPap actuators for haptic application, *Sensors Actuators A: Phys.* **164**(1–2), 68–73 (2010).

27. S. Yun and J. Kim, Multi-walled carbon nanotubes–cellulose paper for a chemical vapor sensor, *Sensors Actuators B: Chem.* **150**(1), 308–313 (2010).

28. S. K. Mahadeva, S. Yun and J. Kim, Flexible humidity and temperature sensor based on cellulose–polypyrrole nanocomposite, *Sensors Actuators A: Phys.* **165**(2), 194–199 (2011).

29. S. Mun, Y. Chen and J. Kim, Cellulose–titanium dioxide–multiwalled carbon nanotube hybrid nanocomposite and its ammonia gas sensing properties at room temperature, *Sensors Actuators B: Chem.* **171**–**172**(0), 1186–1191 (2012).

30. S.-D. Jang, B.-W. Kang and J. Kim, Frequency selective surface based passive wireless sensor for structural health monitoring, *Smart Mater. Struct.* **22**(2), 1–7 (2013).

31. S. K. Mahadeva, H.-U. Ko and J. Kim, Investigation of cellulose and tin oxide hybrid composite as a disposable pH sensor, *Zeitschrift Fur Physikalische Chemie — Int. J. Res. Phys. Chem. Chem. Phys.* **227**(4), 419–428 (2013).

32. J. Kim, H. Lee and H. S. Kim, Vibration sensor characteristics of piezoelectric electro-active paper, *J. Intell. Mater. Syst. Struct.* **21**(11), 1123–1130 (2010).

33. J. Kim, H. Lee and H. S. Kim, Beam vibration control using cellulose-based electro-active paper sensor, *Int. J. Prec. Eng. Manuf.* **11**(6), 823–827 (2010).

34. S. K. Mahadeva and J. Kim, Porous tin-oxide-coated regenerated cellulose as disposable and low-cost alternative transducer for urea detection, *IEEE Sens. J.* **13**(6), 2223–2228 (2013).

35. J. Hu, S. Wang, L. Wang, F. Li, B. Pingguan-Murphy, T. J. Lu and F. Xu, Advances in paper-based point-of-care diagnostics, *Biosens. Bioelectron.* **54**, 585–597 (2014).

36. G. Posthuma-Trumpie, J. Korf and A. van Amerongen, Lateral flow (immuno)assay: Its strengths, weaknesses, opportunities and threats. A literature survey, *Anal. Bioanal. Chem.* **393**(2), 569–582 (2009).

37. D. Mark, S. Haeberle, G. Roth, F. von Stetten and R. Zengerle, Microfluidic lab-on-a-chip platforms: Requirements, characteristics and applications, *Chem. Soc. Rev.* **39**(3), 1153–1182 (2010).

38. Bio-AMD Digital Strip Reader, http://www.bioamd.com/technology.

39. D. D. Liana, B. Raguse, J. J. Gooding and E. Chow, Recent advances in paper-based sensors, *Sensors* **12**(9), 11505–11526 (2012).

40. A. W. Martinez, S. T. Phillips, M. J. Butte and G. M. Whitesides, Patterned paper as a platform for inexpensive, low-volume, portable bioassays, *Angew. Chem. Int. Ed.* **46**(8), 1318–1320 (2007).

41. A. W. Martinez, S. T. Phillips, G. M. Whitesides and E. Carrilho, Diagnostics for the developing world: Microfluidic paper-based analytical devices, *Anal. Chem.* **82**(1), 3–10 (2010).

42. Y. Lu, W. W. Shi, L. Jiang, J. H. Qin and B. C. Lin, Rapid prototyping of paper-based microfluidics with wax for low-cost, portable bioassay, *Electrophoresis* **30**(9), 1497–1500 (2009).

43. T. Songjaroen, W. Dungchai, O. Chailapakul and W. Laiwattanapaisal, Novel, simple and low-cost alternative method for fabrication of paper-based microfluidics by wax dipping, *Talanta* **85**(5), 2587–2593 (2011).

44. D. A. Bruzewicz, M. Reches and G. M. Whitesides, Low-cost printing of poly(dimethylsiloxane) barriers to define microchannels in paper, *Anal. Chem.* **80**(9), 3387–3392 (2008).

45. K. Abe, K. Suzuki and D. Citterio, Inkjet-printed microfluidic multianalyte chemical sensing paper, *Anal. Chem.* **80**(18), 6928–6934 (2008).

46. X. Li, J. Tian, T. Nguyen and W. Shen, Paper-based microfluidic devices by plasma treatment, *Anal. Chem.* **80**(23), 9131–9134 (2008).

47. E. M. Fenton, M. R. Mascarenas, G. P. López and S. S. Sibbett, Multiplex lateral-flow test strips fabricated by two-dimensional shaping, *ACS Appl. Mater. Interfaces* **1**(1), 124–129 (2009).

48. E. Carrilho, A. W. Martinez and G. M. Whitesides, Understanding wax printing: A simple micropatterning process for paper-based microfluidics, *Anal. Chem.* **81**(16), 7091–7095 (2009).

49. M. S. Khan, D. Fon, X. Li, J. Tian, J. Forsythe, G. Garnier and W. Shen, Biosurface engineering through ink jet printing, *Colloids Surf. B: Biointerfaces* **75**(2), 441–447 (2010).

50. X. Li, J. Tian, G. Garnier and W. Shen, Fabrication of paper-based microfluidic sensors by printing, *Colloids Surf. B: Biointerfaces* **76**(2), 564–570 (2010).

51. W. Dungchai, O. Chailapakul and C. S. Henry, A low-cost, simple, and rapid fabrication method for paper-based microfluidics using wax screen-printing, *Analyst* **136**(1), 77–82 (2011).

52. G. Chitnis, Z. Ding, C.-L. Chang, C. A. Savran and B. Ziaie, Laser-treated hydrophobic paper: An inexpensive microfluidic platform, *Lab Chip* **11**(6), 1161–1165 (2011).

53. Y. L. Han, J. Hu, G. M. Genin, T. J. Lu and F. Xu, BioPen: Direct writing of functional materials at the point of care, *Sci. Rep.* **4**, 4782 (2014).

54. B. Kumar, M. Castro and J. F. Feller, Poly(lactic acid)-multi-wall carbon nanotube conductive biopolymer nanocomposite vapour sensors, *Sensors Actuators B: Chem.* **161**(1), 621–628 (2012).

55. C. M. Yu, L. L. Gou, X. H. Zhou, N. Bao and H. Y. Gu, Chitosan-Fe$_3$O$_4$ nanocomposite based electrochemical sensors for the determination of bisphenol A, *Electrochim. Acta* **56**(25), 9056–9063 (2011).

56. T. I. Nasution, I. Nainggolan, S. D. Hutagalung, K. R. Ahmad and Z. A. Ahmad, The sensing mechanism and detection of low concentration acetone using chitosan-based sensors, *Sensors and Actuators B: Chem.* **177**, 522–528 (2013).

57. M. Darder, M. Colilla and E. Ruiz-Hitzky, Chitosan-clay nanocomposites: Application as electrochemical sensors, *Appl. Clay Sci.* **28**(1–4), 199–208 (2005).

58. L. Liu, L. Jin, J. Li, Y. Ran and B.-O. Guan, Fabrication of highly stable microfiber structures via high-substituted hydroxypropyl cellulose coating for device and sensor applications, *Opt. Lett.* **40**(7), 1492–1495 (2015).

59. K. K. Sadasivuni, A. Kafy, L. Zhai, H.-U. Ko, S. Mun and J. Kim, Transparent and flexible cellulose nanocrystal/reduced graphene oxide film for proximity sensing, *Small* **11**(8), 994–1002 (2015).

60. K. Devarayan and B.-S. Kim, Reversible and universal pH sensing cellulose nanofibers for health monitor, *Sensors Actuators B: Chem.* **209**, 281–286 (2015).

61. J. P. Comer, Semiquantitative specific test paper for glucose in urine, *Anal. Chem.* **28**, 1748–1750 (1956).

62. Hydrion Brilliant Plastic pH stript 1–14, http://www.microessentiallab.com/ProductInfo/F60-WIDRG-010140-PST.aspx.

63. H. Miyaji, W. Sato and J. L. Sessler, Naked-eye detection of anions in dichloromethane: Cclorimetric anion sensors based on calix 4 pyrrole, *Angew. Chem. Int. Ed.* **39**(10), 1777–1780 (2000).

64. T. Gunnlaugsson, M. Glynn, G. M. Tocci, P. E. Kruger and F. M. Pfeffer, Anion recognition and sensing in organic and aqueous media using luminescent and colorimetric sensors, *Coord. Chem. Rev.* **250**(23–24), 3094–3117 (2006).

65. R. M. Duke, E. B. Veale, F. M. Pfeffer, P. E. Kruger and T. Gunnlaugsson, Colorimetric and fluorescent anion sensors: An overview of recent developments in the use of 1,8-naphthalimide-based chemosensors, *Chem. Soc. Rev.* **39**(10), 3936–3953 (2010).

66. H. N. Kim, W. X. Ren, J. S. Kim and J. Yoon, Fluorescent and colorimetric sensors for detection of lead, cadmium, and mercury ions, *Chem. Soc. Rev.* **41**(8), 3210–3244 (2012).

67. E. J. Cho, B. J. Ryu, Y. J. Lee and K. C. Nam, Visible colorimetric fluoride ion sensors, *Org. Lett.* **7**(13), 2607–2609 (2005).

68. N. Ratnarathorn, O. Chailapakul, C. S. Henry and W. Dungchai, Simple silver nanoparticle colorimetric sensing for copper by paper-based devices, *Talanta* **99**, 552–557 (2012).

69. M. C. Janzen, J. B. Ponder, D. P. Bailey, C. K. Ingison and K. S. Suslick, Colorimetric sensor Arrays for volatile organic compounds, *Anal. Chem.* **78**(11), 3591–3600 (2006).

70. C. Zhang and K. S. Suslick, A colorimetric sensor array for organics in water, *J. Am. Chem. Soc.* **127**(33), 11548–11549 (2005).

71. J. Halova, O. Strouf, P. Zak, A. Sochozova, N. Uchida, T. Yuzuri and K. Sakakibara, M. Hirota, QSAR of catechol analogs against malignant melanoma using fingerprint descriptors, *Quant. Struct.-Act. Relat.* **17**(1), 37–39 (1998).

72. S. Firestein, How the olfactory system makes sense of scents, *Nature* **413**(6852), 211–218 (2001).

73. H. T. Chueh and J. V. Hatfield, A real-time data acquisition system for a hand-held electronic nose ((HEN)-E-2), *Sensors Actuators B: Chem.* **83**(1–3), 262–269 (2002).

74. B. A. Botre, D. C. Gharpure and A. D. Shaligram, Embedded electronic nose and supporting scftware tool for its parameter optimization, *Sensors Actuators B: Chem.* **146**(2), 453–459 (2010).

75. M. Brattoli, G. de Gennaro, V. de Pinto, A. D. Loiotile, S. Lovascio and M. Penza, Odour detection methods: Olfactometry and chemical sensors, *Sensors* **11**(5), 5290–5322 (2011).

76. K. Heidarbeigi, S. S. Mohtasebi, A. Foroughirad, M. Ghasemi-Varnamkhasti, S. Rafiee and K. Rezaei, Detection of adulteration in saffron samples using electronic nose, *Int. J. Food Prop.* **18**(7), 1391–1401 (2015).

77. L. P. Pathange, P. Mallikarjunan, R. P. Marini, S. O'Keefe and D. Vaughan, Non-destructive evaluation of apple maturity using an electronic nose system, *J. Food Eng.* **77**(4), 1018–1023 (2006).

78. S. Saevels, J. Lammertyn, A. Z. Berna, E. A. Veraverbeke, C. Di Natale and B. M. Nicolai, Electronic nose as a non-destructive tool to evaluate the optimal harvest date of apples, *Postharvest Biol. Technol.* **30**(1), 3–14 (2003).

79. P. Sharma, A. Ghosh, B. Tudu, S. Sabhapondit, B. D. Baruah, P. Tamuly, N. Bhattacharyya and R. Bandyopadhyay, Monitoring the fermentation process of black tea using QCM sensor based electronic nose, *Sensors Actuators B: Chem.* **219**, 146–157 (2015).

80. S. Aathithan, J. C. Plant, A. N. Chaudry and G. L. French, Diagnosis of bacteriuria by detection of volatile organic compounds in urine using an automated headspace analyzer with multiple conducting polymer sensors, *J. Clin. Microbiol.* **39**(7), 2590–2593 (2001).

81. M. Phillips, R. N. Cataneo, A. R. C. Cummin, A. J. Gagliardi, K. Gleeson, J. Greenberg, R. A. Maxfield and W. N. Rom, Detection of lung cancer with volatile markers in the breath, *Chest* **123**, 2115–2123 (2003).

82. M. Phillips, M. Sabas and J. Greenberg, Increased pentane and carbon-disulfide in the breath of patients with schizophrenia, *J. Clin. Pathol.* **46**(9), 861–864 (1993).

83. W. Ping, T. Yi, H. B. Xie and F. R. Shen, A novel method for diabetes diagnosis based on electronic nose, *Biosens. Bioelectron.* **12**(9–10), 1031–1036 (1997).

84. R. E. Baby, M. Cabezas and E. N. W. de Reca, Electronic nose: A useful tool for monitoring environmental contamination, *Sensors Actuators B: Chem.* **69**(3), 214–218 (2000).

85. J. W. Gardner, H. W. Shin, E. L. Hines and C. S. Dow, An electronic nose system for monitoring the quality of potable water, *Sensors Actuators B: Chem.* **69**(3), 336–341 (2000).

86. L. Zhang, F. C. Tian, H. Nie, L. J. Dang, G. R. Li, Q. Ye and C. Kadri, Classification of multiple indoor air contaminants by an electronic nose and a hybrid support vector machine, *Sensors Actuators B: Chem.* **174**, 114–125 (2012).

87. G. Bunte, J. Hurttlen, H. Pontius, K. Hartlieb and H. Krause, Gas phase detection of explosives such as 2,4,6-trinitrotoluene by molecularly imprinted polymers, *Anal. Chim. Acta* **591**(1), 49–56 (2007).

88. Foodsniffer PERES (Foodsniffer (ed.)). http://www.myfoodsniffer.com/ (2014).

89. H. Y. Hsieh and K. T. Tang, VLSI Implementation of a bio-inspired olfactory spiking neural network, *IEEE Trans. Neural Netw. Learn. Syst.* **23**(7), 1065–1073 (2012).

90. Y. S. Kim, S. C. Ha, Y. Yang, Y. J. Kim, S. M. Cho, H. Yang and Y. T. Kim, Portable electronic nose system based on the carbon black-polymer composite sensor array, *Sensors Actuators B: Chem.* **108**(1–2), 285–291 (2005).

91. D. Markham, Portable "electronic nose" smells your meat for you and identifies food poisoning risk. https://www.treehugger.com/gadgets/portable-electronic-nose-smells-your-meat-identifies-food-poisoning-risk.html, Accessed April 17 (2014).
92. Y. G. Vlasov, E. A. Bychkov and A. V. Bratov, Ion-selective field-effect transistor and chalcogenide glass ion-selective electrode systems for biological investigations and industrial applications, *Analyst* **119**(3), 449–454 (1994).
93. A. Legin, A. Rudnitskaya, Y. Vlasov, C. Di Natale, F. Davide and A. D'Amico, A, Tasting of beverages using an electronic tongue, *Sensors Actuators B: Chem.* **44**(1–3), 291–296 (1997).
94. C. Di Natale, R. Paolesse, A. Macagnano, A. Mantini, A. D'Amico, A. Legin, L. Lvova, A. Rudnitskaya and Y. Vlasov, Electronic nose and electronic tongue integration for improved classification of clinical and food samples, *Sensors Actuators B: Chem.* **64**(1–3), 15–21 (2000).
95. G. L. Gil, J. M. Barat, I. Escriche, E. Garcia-Breijo, R. Martinez-Manez and J. Soto, An electronic tongue for fish freshness analysis using a thick-film array of electrodes, *Microchim. Acta* **163**(1–2), 121–129 (2008).
96. F. Winquist, P. Wide and I. Lundstrom, An electronic tongue based on voltammetry, *Anal. Chim. Acta* **357**(1–2), 21–31 (1997).
97. Y. Vlasov, A. Legin and A. Rudnitskaya, Electronic tongues and their analytical application, *Anal. Bioanal. Chem.* **373**(3), 136–146 (2002).
98. C. Di Natale, A. Macagnano, F. Davide, A. D'Amico, A. Legin, Y. Vlasov, A. Rudnitskaya and B. Selezenev, Multicomponent analysis on polluted waters by means of an electronic tongue, *Sensors Actuators B: Chem.* **44**(1–3), 423–428 (1997).
99. C. F. Winquist, S. Holmin, C. Krantz-Rulcker, P. Wide and I. Lundstrom, A hybrid electronic tongue, *Anal. Chim. Acta* **406**(2), 147–157 (2000).

Chapter 5

Biosensors

Gwang-Wook Hong and Joo-Hyung Kim*

Laboratory of Intelligent Devices and Thermal Control
Department of Mechanical Engineering, Inha University
100 Inha-Ro, Nam-Ku, Incheon 22212, Republic of Korea
*joohyung.kim@inha.ac.kr

Biosensors are analytical devices which can convert a biological response into an electrical signal. Started in the 1960s, various types of biosensors have been suggested, studied and commercialized. To design a biosensor, multidisciplinary research in chemistry, biology, and engineering i.e. sensing materials, transducing devices, and immobilization methods is required. This chapter provides basics of biosensors, various types of bio-receptors and transducers. Also disposable biosensors and chemical sensors are discussed. Finally the chapter provides some technical outlooks of disposable biosensors.

1. Basics of Biosensor and Chemical Sensor

1.1. *Introduction*

In order to improve quality of life, simple instruments to easily measure various chemicals are needed in various fields such as health management, environmental conservation, agriculture, and the chemical industries. To ensure a safe working environment and quality in industries, measuring various chemicals quickly and accurately is very important. There are a number of chemicals of importance in the areas of health care, environmental protection, agriculture, and the chemical industry. Therefore, various physico-chemical devices are

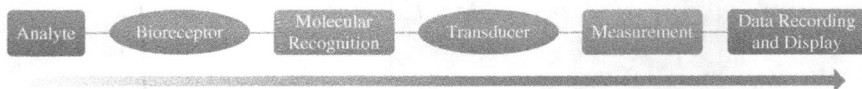

Fig. 5.1. Measurement flow of biosensor. (From an analyte to data display to connecting bioreceptor and transducer.)

employed to measure the concentration or quantity of these chemicals. In general, complex methods and long operational times are required to check for the presence of various harmful chemicals. In place of these measuring devices, the development of direct, quick, and simple monitoring systems is essential.[1]

A biosensor is an analytical device used to detect and measure an analyte by combining a biological component. It consists of a biological recognition system, called a bioreceptor, and a transducer. The interaction of the analyte with the bioreceptor is designed to produce an effect, which is measured by the transducer, which, in turn, converts the information into a measurable effect, such as an electrical signal.

Figure 5.1 shows the basic measurement scheme of a biosensor. A particular analyte can be captured by an appropriate bioreceptor through molecular recognition. The related transducer can produce different signals according to the interaction between the analyte (target molecule) and the bioreceptor. Then the interaction can be displayed or stored in a measurement system. Therefore, biosensors can register a physical, chemical, or biological change and convert such change into a measurable signal. The sensor includes a recognition element that enables its selective response to a particular analyte or a group of analytes, thus minimizing interferences from other sample components.[2]

1.1.1. *Basics of Biosensors and Chemical Sensors*

The sensing element is very important for capturing the target species in order to be analyzed with modern microelectronics and optoelectronics, thereby giving rise to powerful new analytical tools in the field of medicine, food and processing industries.[4] Another main component of a sensor is the transducer or the detector device

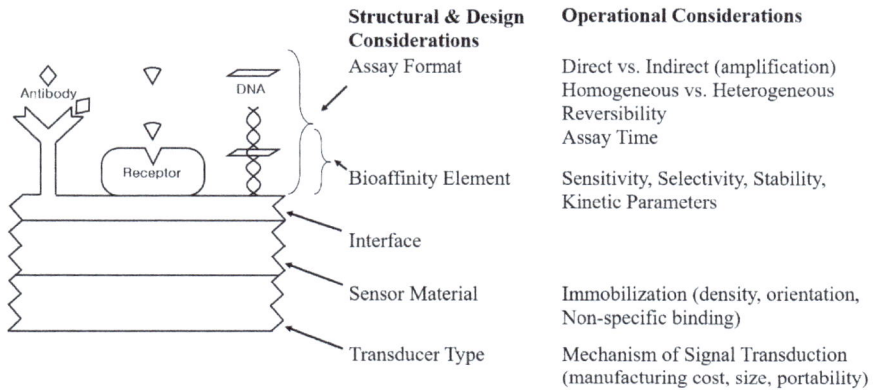

Fig. 5.2. Basic working principle of biosensing.[3]

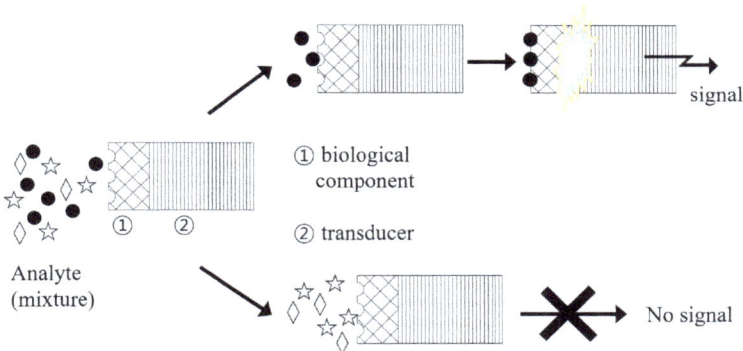

Fig. 5.3. Basic working principle of a biosensor.[4]

that converts the interactions into signals. A signal processor collects, amplifies and displays the signal.[2]

When designing a biosensing structure, several issues such as bioaffinity elements, interfaces, sensing materials including substrate and proper transducers, should be considered. The detailed structure and operational considerations are depicted in Fig. 5.2.[3]

The target molecule (solid circles) from a mixture interacts with the biological part of the sensor, as shown in Fig. 5.3. The biological signal between the target molecules and the biological component is

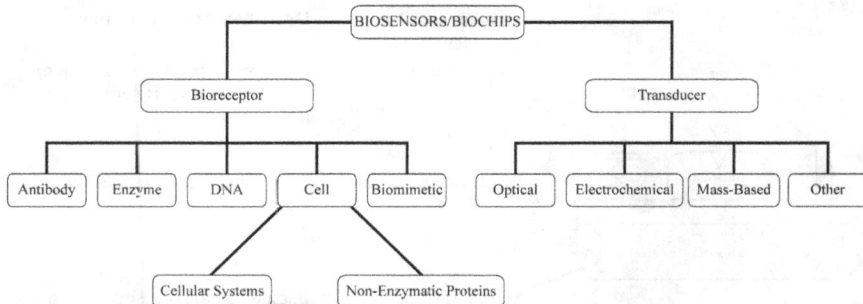

Fig. 5.4.   Classifications of a biosensor.[5]

now converted into a physical signal (e.g., electric or optical) by a transducer. Substances that are not capable of interacting with the biological component will not produce any signal.[4]

Bioreceptors are the key material for specificity in the biosensing mechanism and are involved in the binding of the specific analytes of interest during measurement. These bioreceptors can take many forms and a variety of bioreceptors have been used to detect or monitor different kinds of analytes using biosensors.

In general, as shown in Fig. 5.4, bioreceptors are classified into five types: antibody/antigen, enzymes, nucleic acids/DNA, cellular structures/cells, and biomimetic receptors. As a broad concept, biosensors include sensors that mimic biological systems even without the use of biological materials.

Among biological materials, enzymes have extremely good properties. They can selectively distinguish and react with specific materials called substrates to play a role as an *in vivo* catalyst. Methods to measure a specific chemical using an enzyme have been employed in analytic chemistry. Enzymes have been widely used as a reagent in the fields of medical and food analysis and many improvements including that of complex manipulation have been made.[5]

A chemical sensor obtains selected information about chemical composition in real time. A typical sensor consists of a chemically sensitive layer coupled to a transducer that converts the

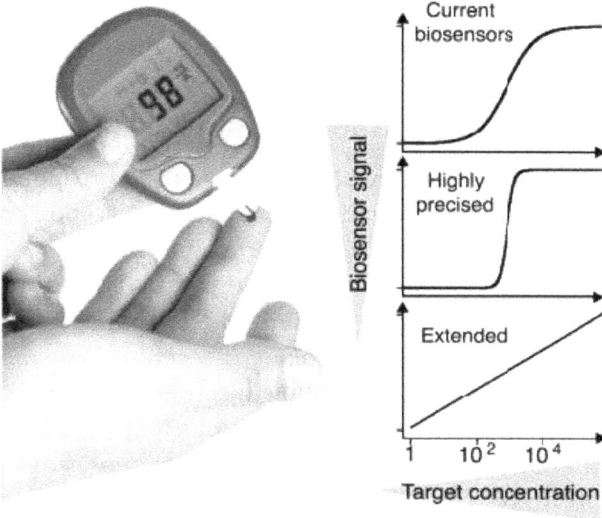

Fig. 5.5. A biosensor for medical diagnostics. Sensor can detect ranges of chemical concentrations.[13]

(bio)chemical interaction into a detectable signal, which is then translated into a digital electronic result. The performance evaluation of a designed biosensor can be done by testing the selection of the target material using an appropriate sensing layer for sensitivity and selectivity. All these substances can be used as a sensing structure, which are combined with various electrodes and transducers. Some examples of biosensors, such as receptor sensors using receptors based on these principles,[6] immune sensors using antigen/antibody reactions,[7−9] cell sensors using animal and plant cells,[10−12] and tissue sensors using animal and plant tissues, have already been developed and some of them are now used in the field of medicine.

Figure 5.5 shows the most popular portable diabetic diagnostic device. The system is composed of transducer/ analytic electronics and testing kit, which can detect the biosignal from the blood. The transducer electronics circuit can be interconnected to a disposable testing kit, which can be used for point-of-care diagnostics. Finally, the analyzed signal is converted into an electric signal for display or transmitted format for recording of the measured data.

2. Receptors for Biosensors and Chemical Sensors

A receptor is similar to a sensory ending that can respond to a stimulus in the internal or external environment. In response to stimuli, the sensory receptor initiates sensory transduction by creating graded potentials or action potentials. For the purposes of biosensing, the receptors should have generic receiving and sending functions.

In biochemistry and pharmacology, a receptor is defined as a protein molecule that receives the chemical signals from outside of a cell, and such type of chemical signals bind to a receptor molecule. These can cause the formation of cellular/tissue response due to change in the electrical activity of the cell. Here, five different types of bioreceptors — antibody/antigen, enzymes, nucleic acids/DNA, cellular structures/cells and biomimetic – are briefly summarized.

Antibody: Antibodies are defined as one of the major classes of proteins constituting about 20% of the total plasma protein and are collectively known as immunoglobulins (Ig). The simplest form of an antibody is defined as the Y-shaped molecule with two identical binding sites for antigens, which are harmful agents. The antigen can be bonded or incorporated into any macromolecule that is capable of inducing an immune response as the antibody has a basic structural unit consisting of four polypeptide chains. The antibody can also act reversibly with a specific antigen. The antibody does not act as a catalyst, and therefore the purpose is to bind other substances.

Enzyme: Enzymes are macromolecular catalysts for biochemical reactions occurring in a particular cell. Due to high enzyme activity, temperature, and pH of the environment, which are also important reaction parameters, enzymes can accelerate or catalyze chemical reactions. Like all catalysts, enzymes increase the rate of a reaction by lowering its activation energy. Some enzymes can make possible the conversion of a substrate into a product many millions of times faster.

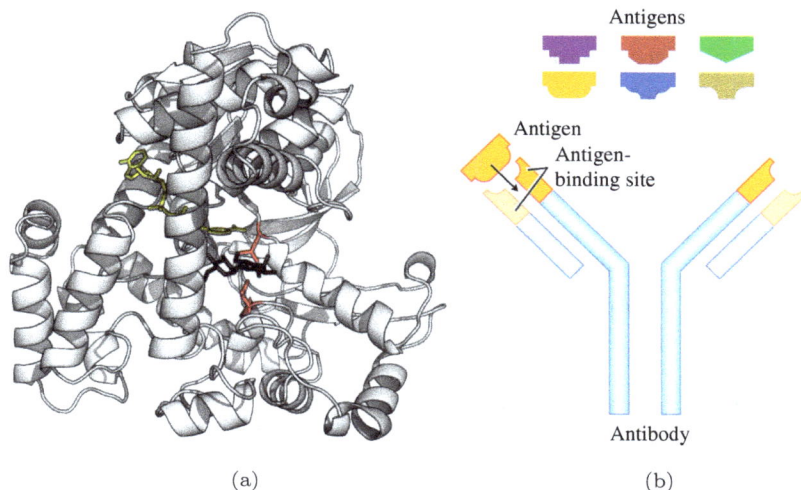

Fig. 5.6. Images of (a) enzyme glucosidase and (b) antibody to bind to a specific antigen. An interaction similar to a lock and key. (Adapted from Wikipedia.)

The molecules at the beginning of the process upon which enzymes may act are called substrates, and the enzyme converts these into different molecules called products. Enzymes, shown in Fig. 5.6(a), are known to catalyze more than 5,000 biochemical reaction types. Most enzymes are proteins, although a few are catalytic RNA molecules. The specificity of enzymes arises from their unique three-dimensional structures. Biomolecules contain a molecular assembly that has the capability of recognizing a target substrate or analyte used in the bioreceptor molecule. Membrane slices or whole cells have also been used in the biosensors. The bioreceptors need a suitable environment for maintaining their structural integrity and biorecognition activity due to signal generated as a result of biorecognition activity in the biosensor.

Protein receptor: Protein molecules have specific affinity for hormones, antibodies, enzymes and other important biologically active compounds. These proteins are mostly bound to the membrane wall. There are hormone receptors, taste receptors for smelling, photoreceptors for eyes, etc. The receptor proteins are responsible for the opening and closing of membrane channels for the transport of specific metabolites. Figure 5.6(b) shows an antibody that binds to a specific antigen.

Nucleic acid/DNA: Nucleic acid is a biopolymer, presented in Fig. 5.7(a), which includes deoxyribonucleic acid (DNA) and ribonucleic acid (RNA). Nucleic acids are formed from monomers known as nucleotides. Each nucleotide has three components: a 5-carbon sugar, a phosphate group, and a nitrogenous base. If the sugar is deoxyribose, the polymer is DNA. If the sugar is ribose, the polymer is RNA. When all three components are combined, they form a nucleotide. Nucleotides are also known as phosphate nucleotides. Nucleic acids are among the most important biological macromolecules (others being amino acids (proteins), sugars (carbohydrates), and lipids (fats)). They are found in abundance in all living things, in which they encode, transmit, and express genetic information. Nucleotides together in a specific sequence provide the mechanism for storing and transmitting hereditary or genetic information via protein synthesis.

Cells (or cellular structures): A cell is the basic structural, functional and biological unit of all living organisms (see Fig. 5.7(b)). Cells consist of cytoplasm enclosed within a membrane, which contains many biomolecules such as proteins and nucleic acids. Organisms can be classified as having a unicellular structure (consisting of a single cell, including bacteria) or a multicellular one (including plants and animals). Most plant and animal cells are visible only under a microscope, with dimensions between 1 and 100 μm.

Biomimetics: Due to the high cost of biorecognition elements based on enzymes, DNA, antibodies, and cells and their low stability, new materials capable of mimicking the response of antibodies have been explored for several decades. A simple low-cost approach for designing selective binding sites in polymeric matrices with molecular imprint technology can results in stable artificial biosensing elements.

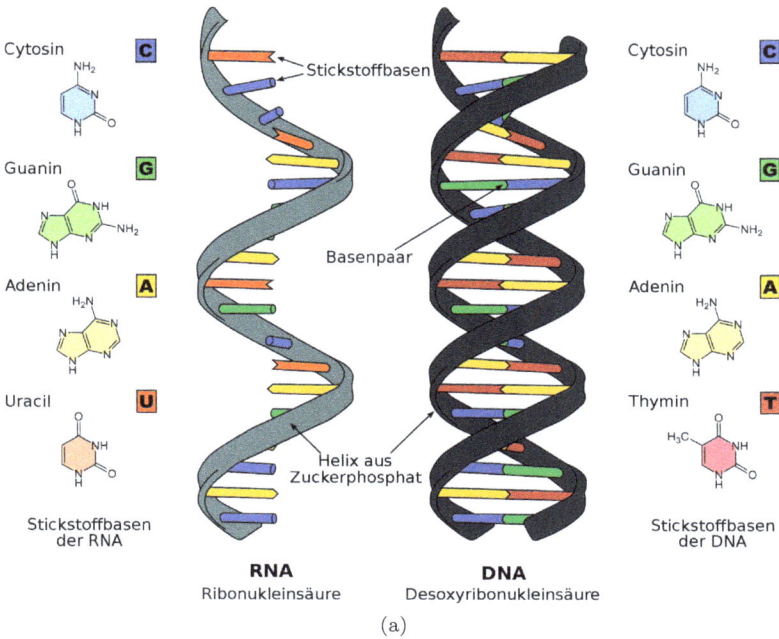

(a)

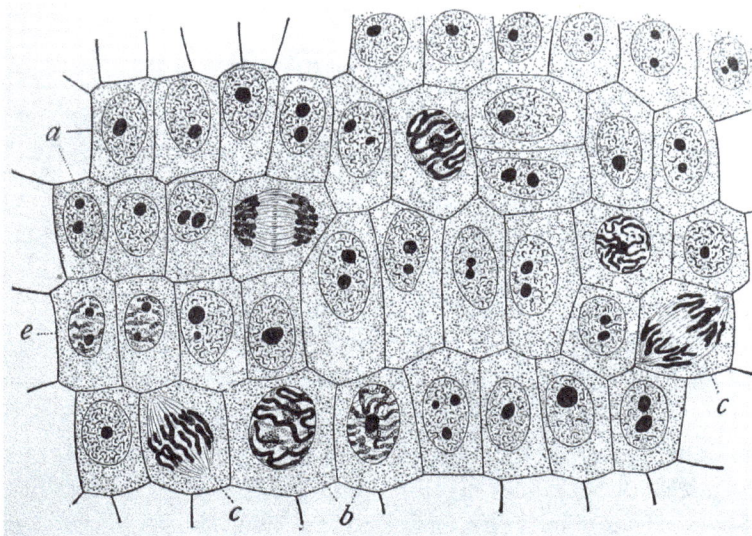

(b)

Fig. 5.7.   Images of (a) nucleic acids and (b) cellular structure of onion. (Adapted from Wikipedia.)

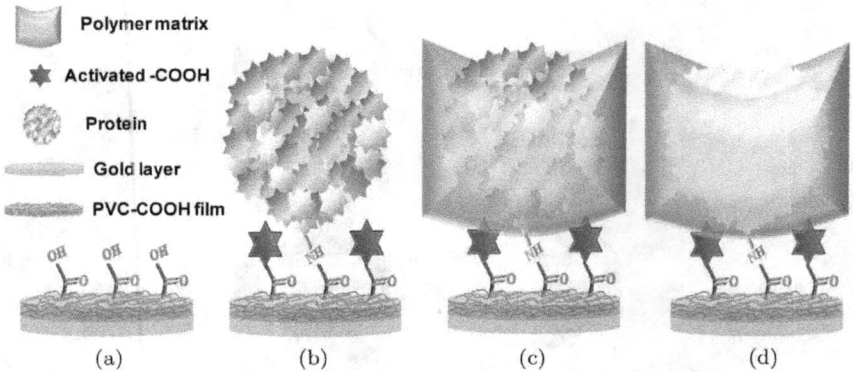

Fig. 5.8. Molecular imprinting polymer based biomimetic based biosensor (a) PVC-COOH film casted the gold surface, (b) subsequent modification with Myo, (c) finally MIP and (d) polymerization step. (Adapted from Wikipedia.)

For example, monitoring Myo in point-of-care was designed by coating the conductive working area of a screen printed electrode with a PVC-COOH film. Figure 5.8 shows the concept of a biomimetic-based biosensor.

3. Transducer Mechanisms of a Biosensor/Chemical Sensor

3.1. Basic Principles of a Transducer for Biosensor

A transducer, coupled receptors, and electrical circuits form important parts of sensing in a biosensor and a chemical sensor. It is the device that converts the physical and chemical interaction into a corresponding electrical signal output. Transducers can be classified by their corresponding output signals from the interaction between the analyte and the receptor of a sensing layer. Typical electrical signals are generated from various physical quantities, such as energy, force, torque, light, motion, position and chemical reaction, etc. Measuring methods of the transducer in biosensing field follow the conventional forms: capacitive, conductometric, conductive, field effect transistor (FET), magneto-elasticity, photometry, piezoelectric and thermometric changes. Figure 5.9 shows the different types of a transducer.

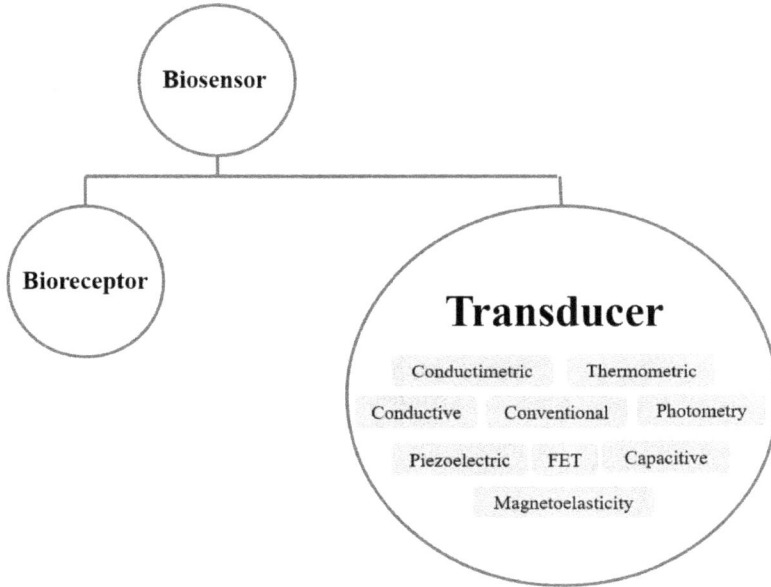

Fig. 5.9. Different types of transducers for biosensor/chemical sensors.

In other words, a transducer can transfer the signal from the output domain of the recognition system to the electrical domain. So a transducer is capable of bi-directional signal transfer (non-electrical to electrical and vice versa). Table 5.1 summarizes the transducer modes according to analytes and receptors.[14]

A biosensor is a self-contained integrated device which is capable of providing specific quantitative or semi-quantitative analytical information using a biological recognition element (biochemical receptor), which is in direct spatial contact with a transducer element. A transducer is used to convert (bio)chemical signal resulting from the interaction of the analyte with the bioreceptor into an electronic one. The intensity of the generated signal is directly or inversely proportional to the analyte concentration. Electrochemical transducers are often used in biosensors. These systems offer advantages such as low cost, simple design, and small dimensions. Biosensors can also be based on gravimetric, calorimetric, or optical

Table 5.1.   Different types of transductors for biosensors and the electrochemical measurements.[14]

Type of analyte	Receptor/chemical recognition system	Measurement technique/ transduction mode
1 Ions	Mixed valence metal oxides Permselective, ion-conductive inorganic crystals, trapped mobile synthetic biological ionophores, ion exchange glasses enzyme(s)	Potentiometric, voltammetric
2 Dissolved gases, vapor, odors	Bilayer lipid or hydrophobic membrane	In series with 1
	Inert metal electrode	Amperometric
	Enzyme(s)	Amperometric or potentiometric
	Antibody, receptor	Amperometric, potentiometric or impedance, piezoelectric, optical
3 Substrates	Enzyme(s)	Amperometric or potentiometric in series with one or two or metal or carbon electrode, conductometric, piezoelectric, optical, calorimetric
	Whole cells Membrane receptors Plant or animal tissue	As above

(*Continued*)

Table 5.1. (*Continued*)

Type of analyte	Receptor/chemical recognition system	Measurement technique/ transduction mode
4 Antibody/antigen	Antigen/antibody oligonucleotide duplex, aptamer	Amperometric, potentiometric or impedimetric, piezoelectric, optical surface plasmon resonance
	Enzyme labeled	In series with 3
	Chemiluminescent or fluorescent labeled	Optical
5 Various proteins and low molecular weight substrates, ions	Specific ligands	As # 4
	Protein receptors and channels	
	Enzyme labeled	
	Fluorescent labeled	

detection.[15] Table 5.2 shows different types of electrochemical transducers.

Potentiometric type: Potentiometric type sensors make use of ion-selective electrodes in order to transduce the biological reaction into an electrical signal. These sensors can be used to determine the analytical concentration of some components of the analyte gas or solution. These sensors also measure the electrical potential of an electrode when no voltage is present. This type of sensor consists of an immobilized enzyme membrane surrounding the probe, typically a pH meter, where the catalyzed reaction generates or absorbs hydrogen ions. The reaction occurring next to the thin sensing glass membrane causes a change in pH, which may be read directly from the pH-meter display. In such electrodes, the electrical potential is

Table 5.2.   Electrochemical transducers classified by type of measurements.[14]

#	Measuring type	Transducer	Transducer analyte
1	Potentiometric	Ion-selective electrode (ISE)	K^+, Cl^-, Ca^{2+}, F^-
		Glass electrode	H^+, Na^+, ...
		Gas electrode	CO_2, NH^3
		Metal electrode	Redox species
2	Amperometric	Metal or carbon electrode	O_2, sugars, alcohols, ...
		chemically modified electrodes (CME)	Sugars, alcohols, phenols, oligonucleotides, ...
3	Conductometric, impedimetric	Interdigitated electrodes, metal electrode	Urea, charged species, oligonucleotides, ...
4	Ion charge or field effect	Ion-sensitive field effect transistor (IS-FET), enzyme FET (en-FET)	H^+, K^+, ...

typically determined at very high impedance, effectively allowing zero current flow and causing no interference with the reaction. Figure 5.10 shows an example of a potentiometric chemical sensor.

Amperometric type: The basic working principle of an amperometric type sensor is the corresponding production of a current when a potential is applied between two electrodes in the sensing process, which is similar to the potentiometric type sensors. As shown in Fig. 5.11, the measurement setup consists of a platinum cathode at which oxygen is reduced and silver/silver chloride is used as a reference electrode. When a potential of -0.6 V relative to the Ag/AgCl electrode is applied to the platinum cathode, a current proportional to the oxygen concentration is produced.

Capacitive type: In a capacitive type sensor, two electrodes are patterned in a metal layer with CMOS circuit architecture and connected to ground. An example of this type biosensor is the one

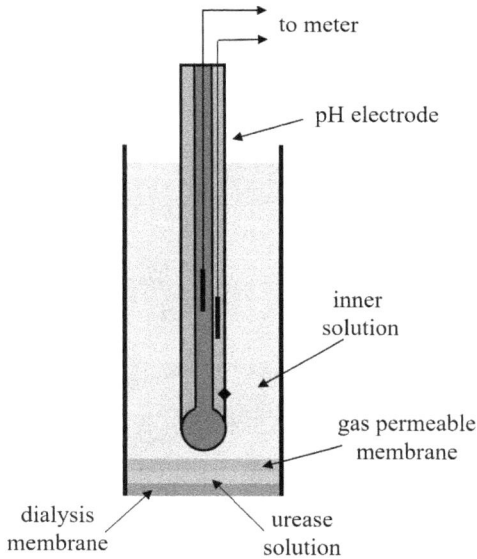

Fig. 5.10. An example of potentiometric type chemical sensor. (Adapted from chem.libretexts.org.)

used for bacteria detection. They are designed to maximize sensitivity to adherent bacteria in the given pixel area, fixed by the underlying readout circuit.[16] This kind of sensor consists of a regular pattern with enough space to let bacteria bind between microsized electrodes, where electric field is maximum. Then, the capacitance of this electrode configuration is impacted mainly by the interaction between the electrode and its counterpart at the ground. Indeed, in the electrical equivalent model, impedance must take into account the insulator capacitance, the dielectric interconnecting the capacitance, the transistor terminal capacitance, and the electrical double layer capacitance in addition to the electrolyte capacitance and resistance. The 2D approximations are given in Fig. 5.12.

When a specific DNA sample is hybridized to oligonucleotides immobilized on the surface of an electrode, water and electrolyte molecules will be displaced by the DNA sample, thereby resulting in

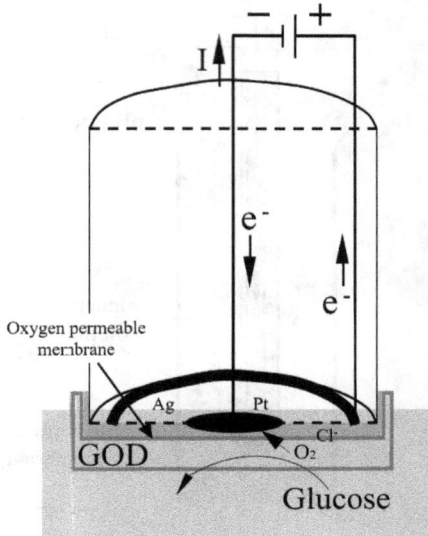

Fig. 5.11. An example of simple amperometric type sensor. (Adapted from www1.lsbu.ac.uk/water/enztech.)

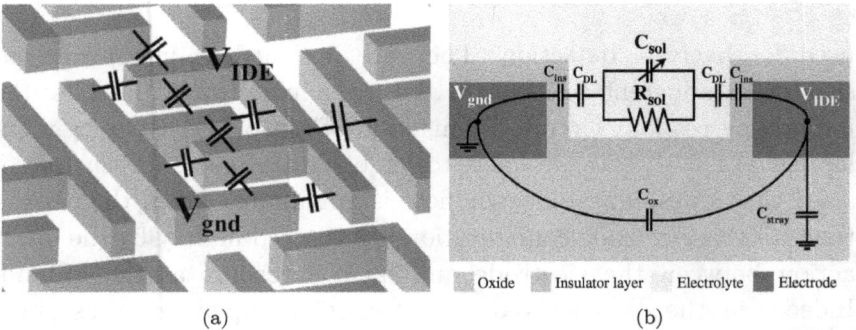

| (a) | (b) |

Fig. 5.12. Capacitive type biosensor. (a) Metal layer electrode of one pixel with the neighbor-less pixels highlighting the concerned capacitance. (b) Cross-section of adjacent electrodes with their equivalent electrical circuit.[16]

a change in capacitance. Figure 5.13 shows a basic concept of DNA detector based on capacitive change.[17]

When a sample containing DNA with a complementary sequence to the oligonucleotide probes is immobilized on the electrode surface,

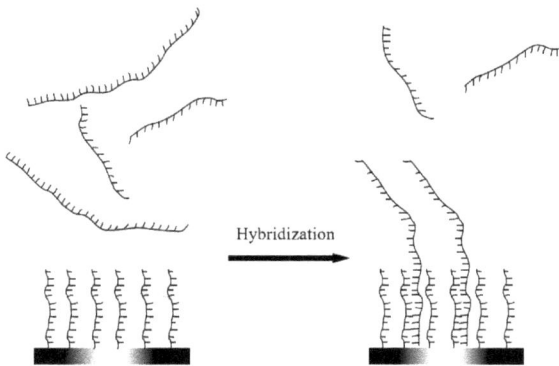

Fig. 5.13.   Detection principle for the capacitive DNA biosensor.[17]

this specific DNA will hybridize on the surface and other DNAs will be eluted in the flow. The hybridized DNA will thereby displace water and solvated ions away from the electrode surface, giving rise to a change in capacitance.

A label-free capacitance-based apta-immunosensor was designed as described in the experimental methods. Interdigital electrode (IDE)-based capacitors were fabricated on SiO_2 wafers in multiple arrays in which each capacitor was designed with 24 gold IDE micro-electrodes in a sensing area of 3 mm^2, which is capable of holding up to 5 μL sample volumes. Here, the capacitive sensor functionalized with anti-VEGF aptamer is called aptasensor.[18] In Fig. 5.14, the aptasensor that was first allowed to capture VEGF protein and later sandwiched with MB-Abs for enhanced signal-to-noise ratio is called "apta-immunosensor," and the stepwise details of experimental methods are described in Fig. 5.14.

Conductometric type: Conductometric sensing is based on the measurement of electrolytic conductivity to monitor the progress of a chemical reaction. As presented in Fig. 5.15, the conductometric transducer is composed of two identical pairs of gold IDE deposited on a ceramic base.[19] To prepare working bioselective membranes, a solution was used. After deposition of the prepared solutions on working surfaces of conductometric transducers, the change

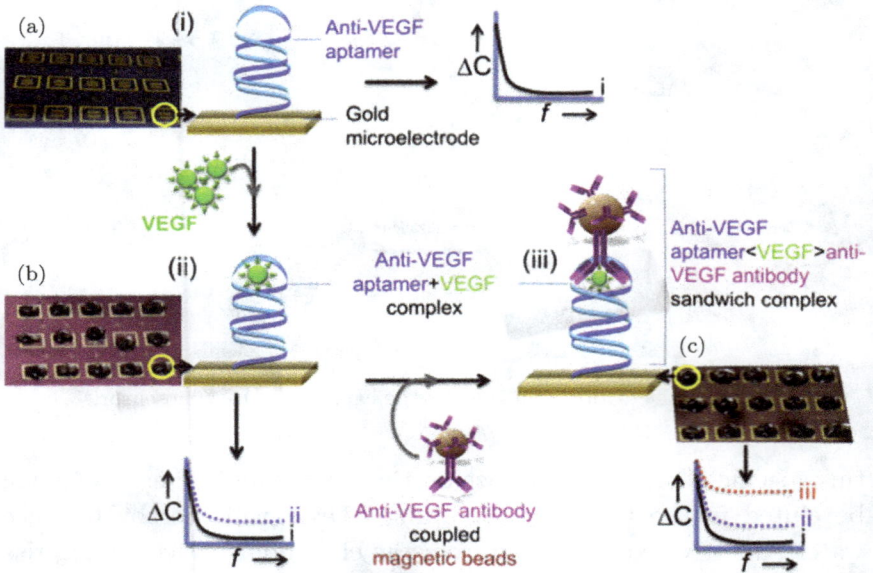

Fig. 5.14.   Schematic illustration of apta-immunosensor showing changes in capacitance against the applied AC electrical frequency: (a) Sensor surfaces functionalized with aptamers (aptasensor) as blank surfaces, (b) aptasensor incubated with 0.1× serum containing spiked VEGF protein at different concentrations and (c) sandwiching of aptasensor with MB-Abs on aptamer–VEGF protein complex formed in step.[18]

of conductance was measured to monitor the chemical reaction. A biosensor analyzer consisting of a sensor block and an electronic measuring block (a portable four-channel biosensor analyzer) was developed in collaboration with the Institute of Electrodynamics of the National Academy of Sciences of Ukraine. The sensor block consisted of a stand with a fixed block of holders; each holder was connected to the fingers of an appropriate conductometric biosensor. An electronic measuring block consisted of the following modules: a secondary transducer and a basic measurement-control module.

Conductive type: In the modified design, the biosensor consists of three disposable membrane pads: sample application pad, capture pad, and absorption pad. The overall biosensor dimension along with the dimensions of each membrane pad has also been tabulated. Silver

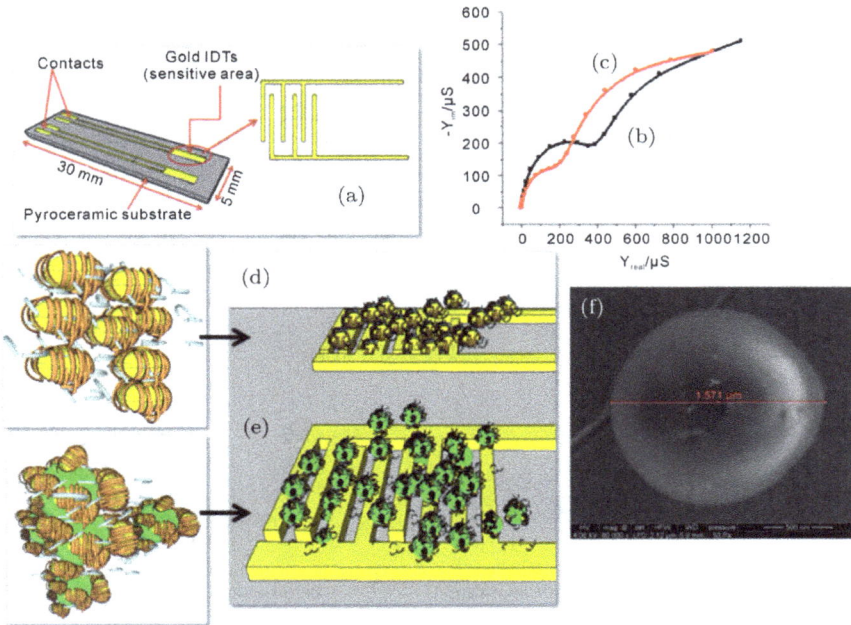

Fig. 5.15. Images of conductometric type biosensor. (a) Scheme of conducto-metric microelectrodes with gold IDTs; (b) complex admittance spectra of bare electrode and (c) modified electrode, (d) BSA (20%) mixed with PAH-coated gold nanoparticles, immobilized on the reference electrode, through cross-linking under glutaraldehyde vapors; (e) Spirulina cells mixed with PAH-coated gold nanoparticles and BSA (20%), immobilized on the working electrode, through cross-linking under glutaraldehyde vapors; (f) SEM image of Spirulina cell.[19]

electrodes were fabricated along both sides on the capture pad, leaving an electrode gap of 0.5 mm. For data acquisition, the biosensor units were connected to a handheld multimeter linked with a computer. This arrangement shows the schematic representation of the immunomagnetic separation and biosensor detection procedure. The biosensor detection involves a sandwich immunoassay, with a capture antibody (IgG) immobilized on the biosensor capture pad and a detector antibody conjugated with synthesized EAPM nanoparticles. The detector antibody-conjugated EAPM nanoparticles are added to the *B. anthracis* spore-contaminated food samples (step 1) and used to immunomagnetically concentrate the spores from

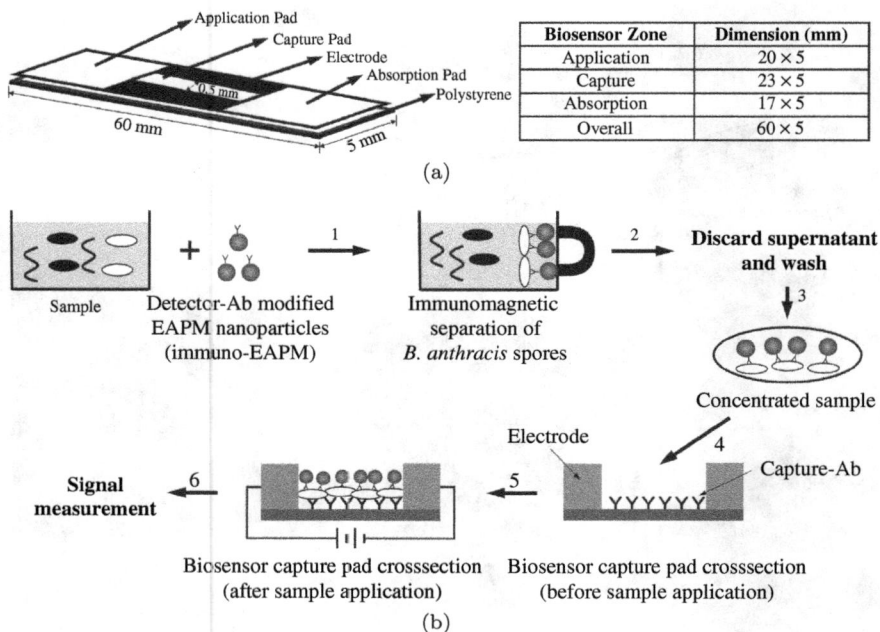

Fig. 5.16. (a) Schematic of conductive nanoparticle-based biosensor architecture and (b) working principle of conductive transducer for the detection system.[20]

the complex food matrices (step 2). The concentrated targets are then washed to remove unbound materials (step 3) and applied to the sample application pad of the biosensor (step 4). The antigen–antibody–EAPM complex flows to the capture pad, where the antigen is anchored by the capture antibodies present and a sandwich complex is formed (step 5). The conductive EAPM nanoparticles bound to the antigens in the sandwich act as a voltage controlled "ON" switch, resulting in decreased resistance, which is recorded electrically across the silver electrodes (step 6). The schematic processes are depicted in Fig. 5.16.

Bio-FET type: Bio-field effect transistor (bio-FET) is a FET gated by biological molecules binding to the FET gate. The detailed structure of bio-FET is shown in Fig. 5.17. Generally, bio-FETs couple a semiconductor device to a biosensitive layer that detects

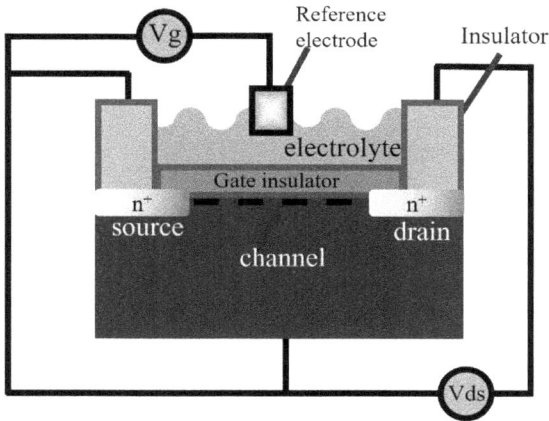

Fig. 5.17. Schematic diagram of FET-based transducer for biosensing. (Adapted from Wikipedia.)

biomolecules such as nucleic acids. A bio-FET system consists of a semiconducting transducer separated by an insulator layer from the receptors, which are the biological recognition element. A bio-FET also consists of a transducer immobilized on the surface containing the target molecule called the analyte. Once the analyte binds to the recognition element, the charge distribution at the surface will change, triggering a change in the electrostatic potential of the semiconductor. The transistor controls the electron flow in the channel using two electrodes, termed source and drain. As a result, the conductance of the semiconductor changes, which can be measured. This change of conductance directly yields the value of concentration. The current flow between the source and drain is controlled by the gate.

Figure 5.18 shows the layout of a EGFET type biosensor. It consists of two parts: a sensor part for signal generation and a transducer part for signal amplification and readout.[21] The active sensing surface is a thin gold film evaporated on a Si/SiO_2 substrate. The gold film is in contact with the electrolyte, and its surface potential depends on the amount of charged molecules attached to the surface. The gold surface can be chemically modified to achieve specific adsorption of certain biomolecules. To measure the potential changes on the gold

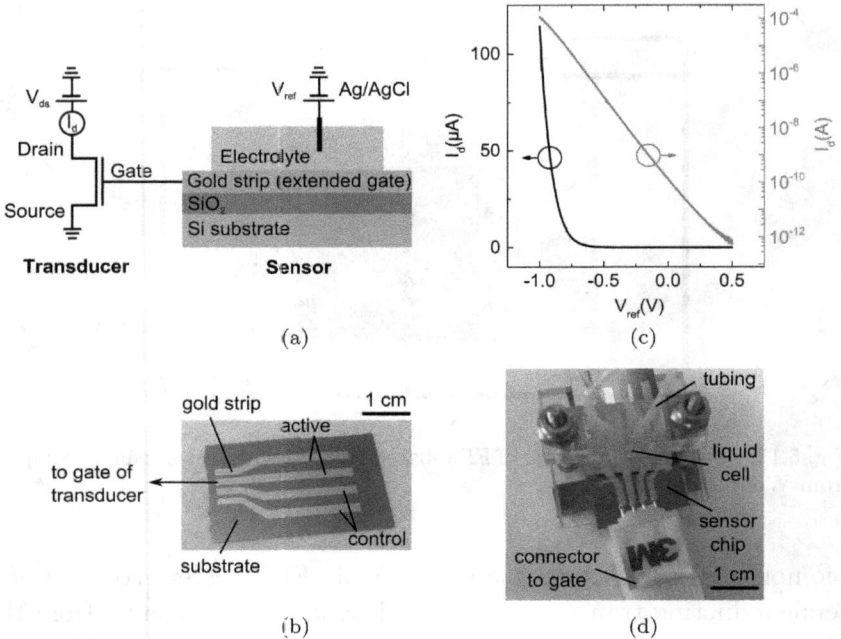

(a)

(c)

(b)

(d)

Fig. 5.18.   EGFET biosensor (a) Schematic of the potentiometric sensing device. (b) A typical transistor transfer curve showing that the drain current Id can be modulated by sweeping the voltage at the reference electrode. (c) Image of a disposable sensing chip. To read out the potential changes generated on the sensor surface, the individual gold strips were connected using a test clip shown in (d) to the gate of the transducer.[21]

surface, a commercial MOSFET is used as the readout transducer. The gold film acts as the extended gate of the MOSFET and is electrically connected to the gate terminal.

Magnetoelastic type: The phenomenon of magnetoelasticity occurs when an external magnetic field is applied to a material. Then the material experiences a shape change due to the superposition of its internal magnetic moments by spin–orbital coupling. If this magnetic field is of an alternating type and the piece of material is a planar structure, then fundamental resonance frequency of the longitudinal vibration is caused by magnetostriction.[22] Then the fundamental resonance frequency is

$$f_0 = \sqrt{\frac{E}{\rho(1 - \nu)}} \frac{1}{2L},$$ (5.1)

where E, ρ, ν, and L are the elastic modulus, density, Poisson's ratio, and length, respectively. Additionally, this resonance frequency will be affected by the sensor's mass load as shown below.

$$\Delta f = \frac{f_0}{2} \frac{\Delta m}{M},$$ (5.2)

where Δf is the change in resonance frequency, Δm is the change in the mass due to the attached load on sensing layer, and M is the original mass of the sensor. Also, this relationship can be used to analyze an attached load that is evenly distributed over the surface of the sensor. Using this basic concept, the resonance frequency shift can be utilized as a biosensing platform as schematically shown in Fig. 5.19. First, the magnetoelastic particle (MEP) is coated with a gold layer to enable biocompatibility with protein molecules such that they may be attached onto the surface. Next, a biomolecular recognition element (either bacteriophage or antibody) is immobilized onto the gold surface, enabling the capture of a single targeted pathogenic species. At this point, the fundamental resonance frequency is measured in order to completely characterize the biosensor prior to exposure to the target containing solutions. This is accomplished by applying an AC magnetic field to set up resonance, as well as a DC biasing magnetic field for amplification of the signal. Then, when the sensor is exposed to the targeted bacteria or spores, they bind to the sensor, thereby increasing its mass and lowering its resonance frequency. Also, based on the magnitude of the frequency shift, the approximate number of attached spores or cells may be calculated (assuming an approximate mass of each spore or cell).

Photometric type: The basic principle of the photometric type of biosensor is measurement of change of light intensity due to interaction with the analyte. In the photochemical oxidation method, short wavelength UV-C (<280 nm) light is commonly employed in a photometric type biosensor. On the other hand, photocatalytic

(a)

(b)

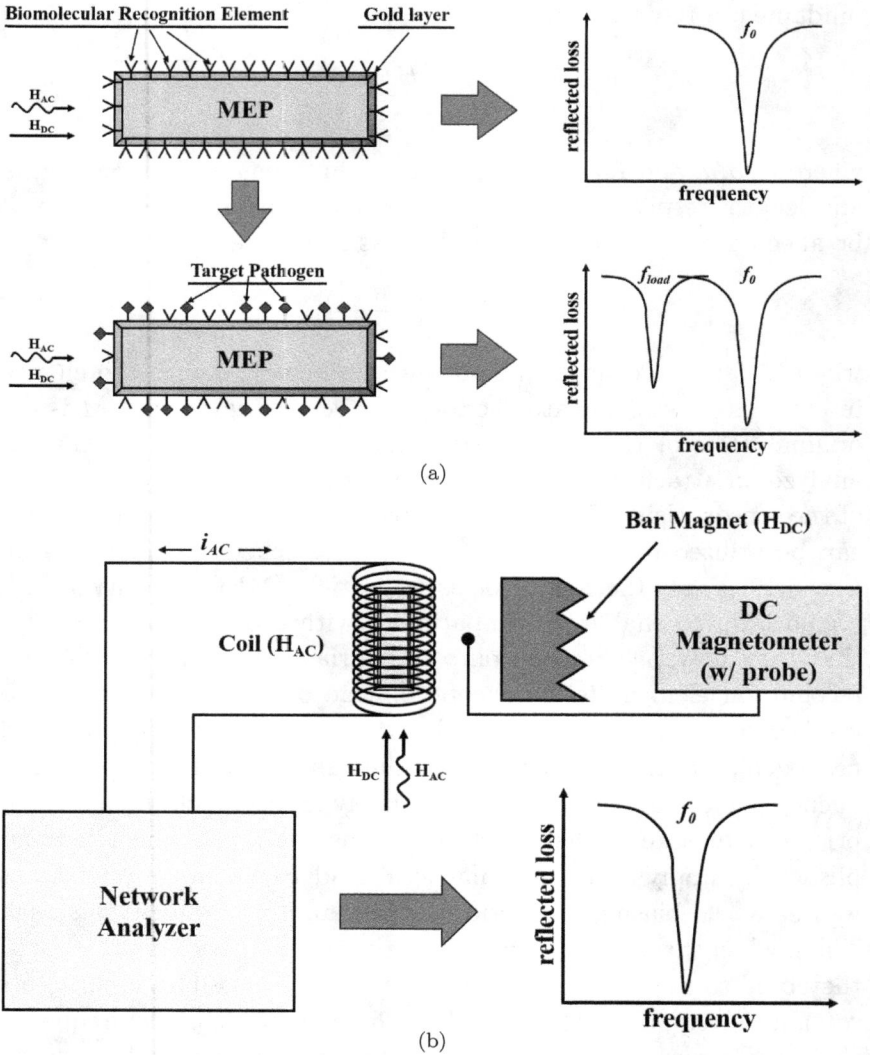

Fig. 5.19. Schematic diagram of the operating principle of a magnetoelastic biosensor. (a) HAC and HDC are the alternating and biasing magnetic fields, The resonance frequency after attachment of the pathogen is f_{load}, such that $\Delta f = f_0 - f_{load}$. (b) Characterization method for the determination of resonance frequency.[22]

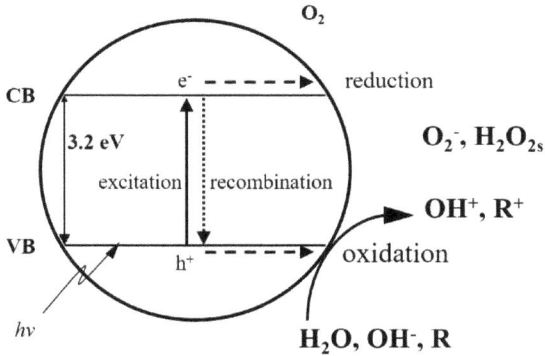

Fig. 5.20. Schematic diagram of simplified mechanism for the photometric sensing by activation of titanium dioxide particle.[23]

experiments with titanium dioxide usually use long wavelength UV-A (>315 nm) light. Generally, the degradation of organic compound in aqueous solution is characterized by electron–hole formation at the surface of an illuminated titanium dioxide particle. When titanium dioxide is illuminated with band gap energy higher than 3.2 eV (380 nm), a photon excites an electron from the valence band (VB) to the conduction band (CB) and leaves an electronic vacancy, commonly referred to as a hole, in the VB. The electron in the CB can be transferred to adsorbed H^+, O_2, or the chlorinated pollutant initiating various reactions. The hole in the VB can react with surface-bound water, hydroxide group, anions, and organic substrate. Therefore, organic compounds in aqueous solution are decomposed by forming lower molecular weight compounds by photocatalytic oxidation. The results would increase the sensitivity of the BOD sensor, because organic compounds of lower molecular weight are more readily biodegradable through the biofilm. Figure 5.20 shows the basic mechanism of the photometric type sensor.

Optical type: In optical biosensors, the biological sensing element is connected to an optical transducer system and the signal is obtained based on absorption, luminescence, or reflectance. Generally, optical biosensors can be divided into two main groups: direct optical detection and indirect optical detection. Biosensors based on surface

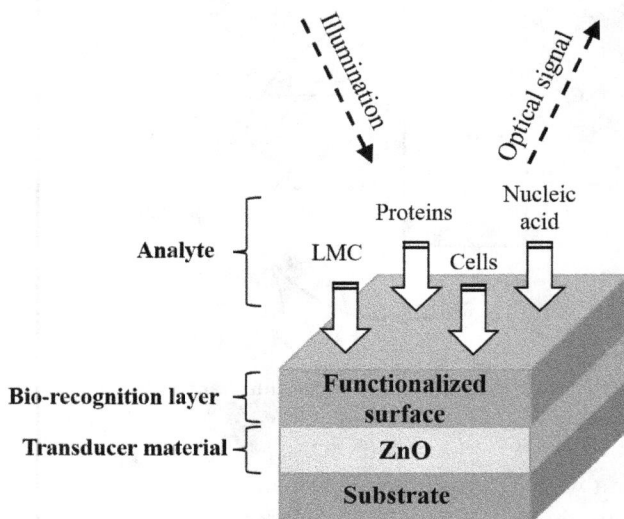

Fig. 5.21. Basic schematic of optical type biosensor.[24]

plasmon resonance (SPR), ellipsometry, reflectometry, interferometry, or photoluminescence are used in direct detection systems. Indirect biosensors (labeled systems) are based on fluorescence, surface-enhanced Raman spectroscopy (SERS), photoluminescence quantum dots (used as labels), and others. The difference between these groups is that in the case of direct optical detection, one measures the properties of the transducer. In the second case, the attached tags (such as fluorescent dye) are used to detect the target analyte. The biosensitive layer is formed by immobilization of the biological recognition element (enzyme, receptor protein, probe molecule, cellreceptor, etc.) on the surface of the transducer. This biorecognition layer (functionalized surface) serves as a basis for capturing the target analyte (e.g. LMC, protein and nucleic acid or cell). The main goal of the optical type biosensor is to achieve precise and rapid detection of the target biomolecules with high sensitivity and selectivity. The general scheme of an optical type biosensor is shown in Fig. 5.21. The light from the source reaches the surface and then optical signal (luminescence or reflectance) is recorded before and after immobilization of

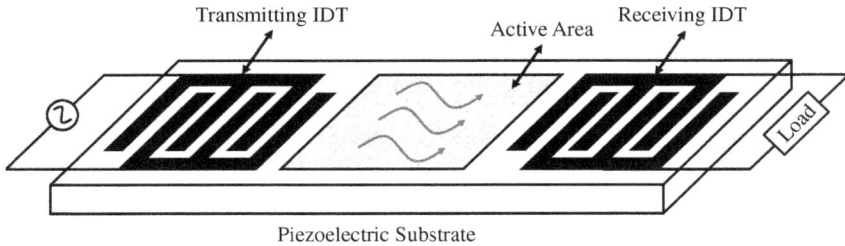

Fig. 5.22. Schematic view of a delay line of SAW-based sensor.[25]

the biosensitive layer, as well as after its interaction with analytes. The changes in optical signal, resulting from the adsorption of the target analyte, allow us to plot a curve showing the dependence of the biosensor signal on analyte concentration.

Piezoelectric type: Surface acoustic wave (SAW) devices, which make use of a piezoelectric substrate, are preferred for high-performance biosensing devices with non-complicated fabricating and detecting procedures due to their excitation techniques. SAW devices use an interdigitated transducer (IDT) consisting of thin metal electrodes coated on piezoelectric materials with a specific equal spacing and width to transmit and receive acoustic waves. Generally, surface acoustic waves travel along the surface of an elastic media. Acoustic waves can be easily generated within a piezoelectric material through the transduction of an applied electric field. The target biomolecules can be captured on the active layer, which is coated with a sensing polymer or structure. The captured biomolecules can increase the mass of the active layer where a significant change occurs in the amplitude and velocity of the propagating wave, resulting in the decrease of the operating frequency of the SAW device. This frequency shift is detected by receiving IDT, which is connected to a frequency counter or network analyzer. Figure 5.22 shows the basic schematic of SAW device used in a biosensing application.[25]

Generally, SAW biosensor consists of two IDTs: the transmitting IDT that launches a mechanical acoustic wave through the piezoelectric substrate and the receiving one that receives the mechanical wave

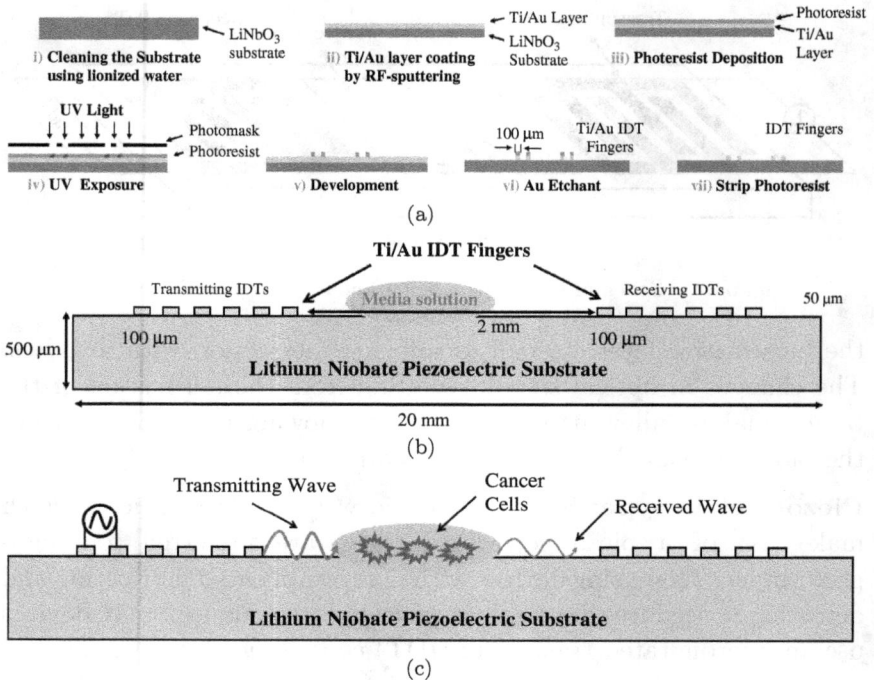

i) **Cleaning the Substrate using Iionized water** — LiNbO$_3$ substrate

ii) **Ti/Au layer coating by RF-sputtering** — Ti/Au Layer / LiNbO$_3$ Substrate

iii) **Photeresist Deposition** — Photoresist / Ti/Au Layer

iv) **UV Exposure** — UV Light / Photomask / Photoresist

v) **Development**

vi) **Au Etchant** — 100 μm / Ti/Au IDT Fingers

vii) **Strip Photoresist** — IDT Fingers

(a)

Ti/Au IDT Fingers

Transmitting IDTs — Media solution — Receiving IDTs — 50 μm

500 μm — 100 μm — 2 mm — 100 μm

Lithium Niobate Piezoelectric Substrate

20 mm

(b)

Transmitting Wave — Cancer Cells — Received Wave

Lithium Niobate Piezoelectric Substrate

(c)

Fig. 5.23. Fabrication steps of SAW biosensor: (a) Schematic view of the cross-section of the SAW biosensor (b) without, and (c) with the attachment of cancer cells.[25]

after a delay of a few moments and transforms it into an electrical signal. The amount of delay is determined by the properties of the medium placed on the active area of the device. Figure 5.23 shows the detailed fabrication process of a SAW sensor for bioapplications.

Another acoustic biosensor type is a direct label-free immunosensor in which a quartz crystal microbalance (QCM) transducer is used for detection. After obtaining the steady-state condition of a QCM, a certain concentration of plasmin prepared in a buffer solution was allowed to flow through the cell and changes in resonant frequency (f_s) and motional resistance (R_m) were continuously and simultaneously measured. Figure 5.24 show the schematic diagram of a QCM-based biosensor.[26]

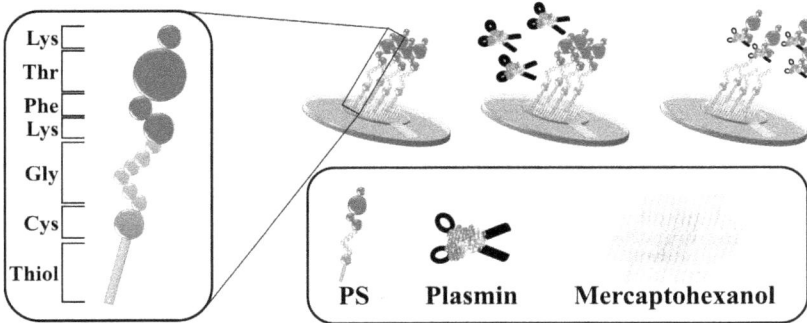

Fig. 5.24. Schematic illustration of the QCM electrode composed of specific PS and mercaptohexanol (MCH) before and after its interaction with plasmin. The scissors representatively cut the PS realizing the short fragment of the peptide cleaved by plasmin.[26]

Thermometric type: Thermometric biosensors exploit the fundamental property of biological reactions, i.e., absorption or evolution of heat, which is reflected as a change in the temperature within the reaction medium. Calorimetry, in which the change in the quantity of heat exchanged is measured, was directly used to calculate the extent of reaction (for catalysis) or structural dynamics of biomolecules in the dissolved state. Calorimetric devices for routine use were limited by the cost of operation and relatively long experimental procedures. However, as presented in Fig. 5.25, the enzyme thermistor is used, which works based on flow injection analysis in combination with an immobilized biocatalyst and heat-sensing element.

The evolution of heat is a general property accompanying biochemical transformations. The total heat evolved is proportional to the molar enthalpy change and the total number of moles of product molecules created in the reaction:

$$Q = n_P(\Delta H), \qquad (5.3)$$

where Q is the total heat, n_P is moles of product, and ΔH is the molar-based enthalpy change. The resulting temperature change (ΔT) is dependent on the total heat capacity of the system including

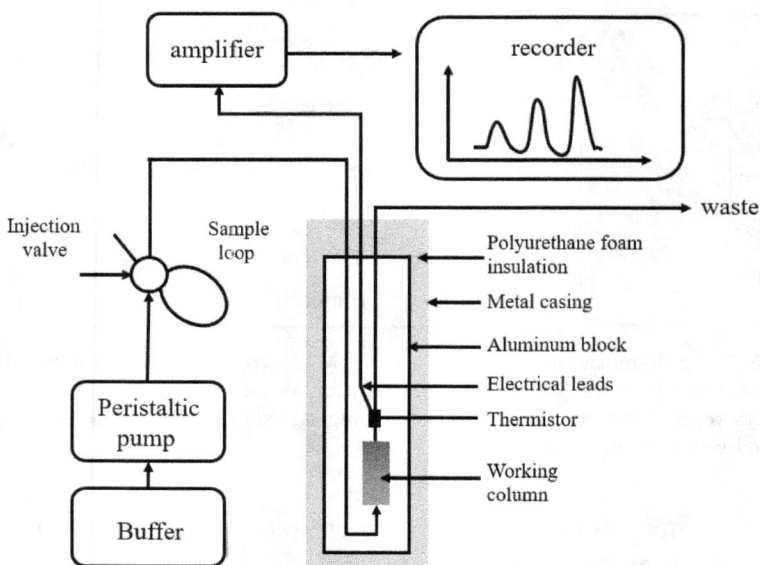

Fig. 5.25. Schematic of enzyme thermistor-based biosensor.[27]

the heat capacity of the solvent. Thus the change in temperature recorded by the thermal biosensors can be defined by

$$\Delta T = -\frac{(n_P \Delta H)}{C_s}. \tag{5.4}$$

The enthalpy changes for enzymatic catalysis is around -10 to -200 kJ/mol, which is adequate for the determination of the substrate concentrations at clinically interesting levels for a range of metabolites including free and esterified cholesterol, ethanol, glucose, lactate, oxalate, triglycerides, and urea. Table 5.3 shows the molar base enthalpy change values of representative enzymes.

4. Disposable Biosensors/Chemical Sensors

Due to high manufacturing costs, cheap and mobile devices with light and flexible substrates, e.g., plastic or paper-like materials, are suggested and developed for use in biosensors and chemical sensors. Paper, composed of multiple layers of cellulose fibers, which can

Table 5.3. Molar enthalpy change values for some common enzyme reactions.[27]

Enzyme	Substrate	$-\Delta H$ (kJ mol^{-1})
Catalase	Hydrogen peroxide	100
Cholesterol oxidase	Cholesterol	53
Glucose oxidase	Glucose	80–100
Hexokinase	Glucose	75[a]
Lactate dehydrogenase	Sodium pyruvate	62
NADH dehydrogenase	NADH	225
β-Lactamase	Penicillin-G	115[a]
Trypsin	Benzoyl-L-arginineamide	29
Urease	Urea	61
Uricase	Urate	49

[a]In Tris buffer (protonation enthalpy -47.5 kJ mol^{-1}).

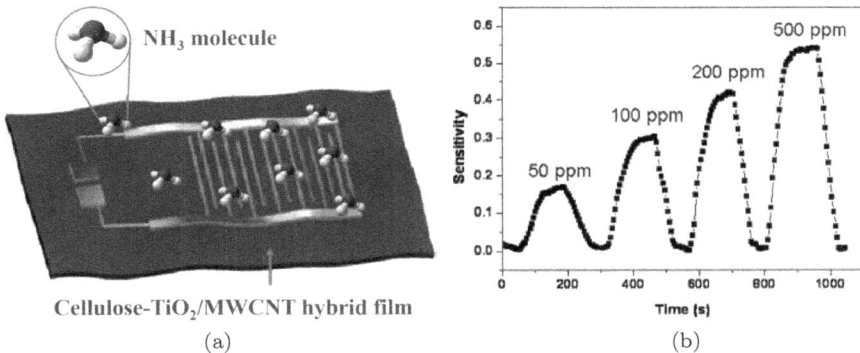

(a)   (b)

Fig. 5.26. Gas sensor based on cellulose composite substrate for detecting NH$_3$.[28]

form an inner network structure, is a flexible, lightweight, cheap, and disposable substrate. If simple printed electronics with sensing structure are fabricated on paper, it will bring in a broad technological impact. Recently, a new approach for semiconducting logic devices based on cellulose paper was reported.[28] Cellulose–SnO-based biosensors, TiO$_2$–cellulose composite for gas monitoring, and GaN–cellulose composite have already been suggested for disposable sensor applications. Figure 5.26 shows the NH$_3$ gas sensing chemical sensor.

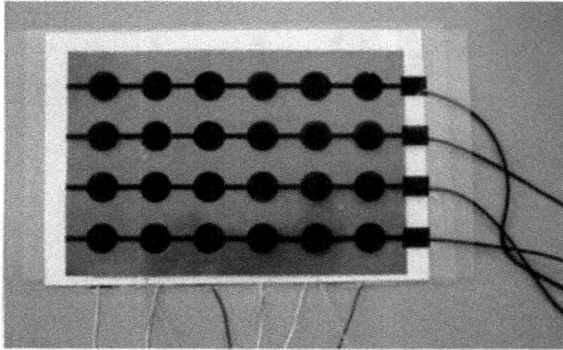

Fig. 5.27. Addressable electrode array-based disposable biosensor.[30]

In recent times, the development of cheap and disposable biosensors has increased significantly for diagnostics in medical application.[29] Figure 5.27 shows the printed electrodes on paper for biosensing applications developed by Nanobiosensors Co. Bacterial cellulose (BC) nanopaper is a multifunctional material known for numerous desirable properties: sustainability, biocompatibility, biodegradability, optical transparency, thermal properties, flexibility, high mechanical strength, hydrophilicity, high porosity, broad chemical-modification capabilities, and high surface area. A novel addressable electrode array based on paper was assembled on the crossing points of the row/column electrodes to form a 4 × 6 sensor array, with one paper layer containing the sensing sites and the other paper layer the printed counter electrode and reference electrode. By combination with nanomaterials, they bring important advantages in the design of novel biosensing systems or improvements in the existing devices. Application of nanomaterials in environmental monitoring (for example, detecting heavy metals) and nanoparticles used for detecting DNA, proteins and even cells (for example, cancer diagnostics) are showing a great potential in enhancing biosensor sensitivity, stability and cost efficiency of the devices.

The most important part of a lateral flow immunoassay (LFIA) is the detection membrane made of cellulose nitrate or nitrocellulose (NC), a porous material where the capture reagents (e.g., antibodies) are immobilized due to a possible combination of electrostatic and hydrophobic forces. Wax printing on membrane is a simple and low-cost patterning technique based on the melting of solid wax, which has been used for fabricating paper-based microfluidics in the NC membrane. From Fig. 5.28, it is obvious that the enhancement of the sensitivity of lateral flow immunoanalysis based on paper (paper-LFIA) platforms is strongly related to the sensitivity of the labels. Paper-LFIAs based on the use of AuNPs not only as labels but also as carriers of enzymatic labels have been developed.[30,31]

Various nanopaper-based optical sensing platforms have been suggested. These platforms can be tuned, using nanomaterials, to exhibit plasmonic or photoluminescent properties that can be exploited for sensing applications (Fig. 5.29). The platforms include a colorimetric-based sensor based on a nanopaper containing embedded silver and gold nanoparticles; a photoluminescent-based sensor, comprising CdSe at ZnS quantum dots conjugated to nanopaper; and a potential up-conversion sensing platform constructed from nanopaper functionalized with $NaYF_4:Yb^{3+}$ at Er^{3+} and SiO_2 nanoparticles. Cellulose-based disposable substrate can provide an advantageous preconcentration platform that facilitates the analysis of small volumes of optically active materials (~ 4 μL). It is suggested that these platforms will pave the way for optical (bio)sensors or theranostic devices that are simple, transparent, flexible, disposable, lightweight, miniaturized, and perhaps wearable.[32]

Recently, the population density and size of silver nanoparticles embedded within bacterial cellulose were shown to be readily modulated by exposure to volatile compounds, leading to a simple naked-eye detection of a hazardous corrosive vapor (ammonia) or a mixture of volatile compounds released during food spoilage using a piece of plasmonic nanopaper. Thus, using a plasmonic nanopaper, we can monitor the color change from amber to light amber upon exposure to ammonia vapor, and from amber to a gray or taupe upon exposure to fish or meat spoilage. Figure 5.30 shows a simple visual detector of NH_3 gas from food for monitoring color change.[33]

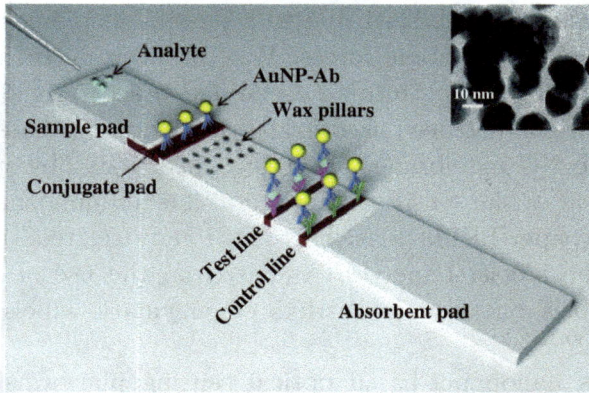

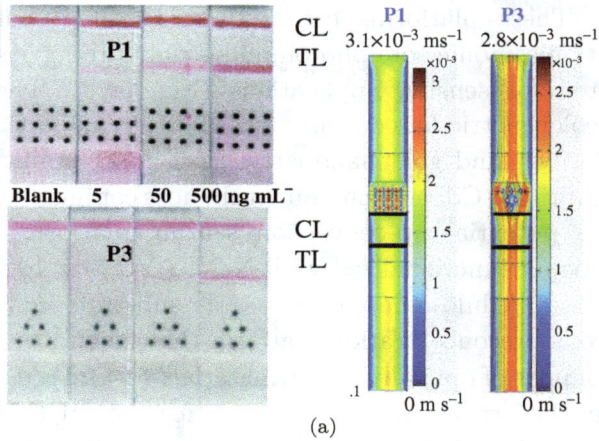

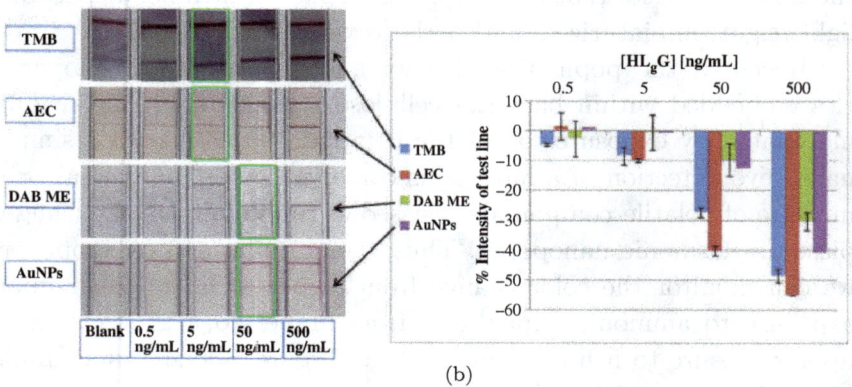

Fig. 5.28. (a) LFIA strip. LF strip with different pillar patterns printed and results. (b) Photos of LFIA strip for different concentrations of HI_gG.[30,31]

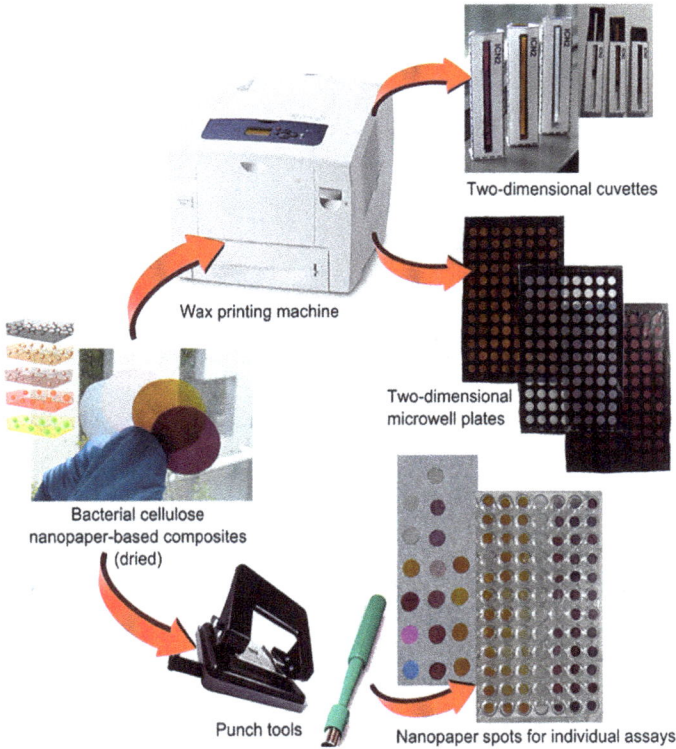

Fig. 5.29. Nanopaper-based optical sensing.[32]

5. Limitations of Biosensors and Outlook for the Future

5.1. *Limitation of Biosensor and Suggestions*

Biosensors have limitations as a device such as limited shelf-life and stability of the biorecognition component as well as nonspecific binding. The modeling was performed taking into account the internal diffusion limitation and a steady-state condition. Due to limitations in diffusion, the apparent stability of biosensor response increased many times in comparison to the stability of the most labile enzyme of the chain. When we use a biosensor, it is possible to measure the appropriate protein concentration. When the

Fig. 5.30.   Visual detector of VOC gas (NH$_3$) using a plasmonic nanopaper.[33]

protein binds to a proper ligand immobilized on the sensor surface, it is completely limited by diffusion and mass transport
phenomena.

For determining protein concentrations, most devices should
measure the whole concentration of the chemical processes not dis-
tinguishing the active from inactive molecules. However, such measurement is not accurate if the sample is not purified and if problems
arise from the limitation of devices that cannot make a distinction
between biologically active and inactive molecules. Fortunately, solutions are available to solve these limitations. Both these limitations
can be overcome by determining protein concentration with biosensor
instruments that make use of surface plasmonic resonance. If mass
transport of the analyte from the bulk flow to the sensing surface
area with immobilized ligand is much faster than the rate of binding
to the ligand, the concentration of the analyte at the sensing surface
will be the same as that in the bulk flow. Therefore, measurement of
analyte kinetics will exclusively reflect the binding kinetics. On the

other hand, when mass transport shows rate-limited behavior, the obtained sensorgrams will be correlated to the mass transport, which depends on active analyte concentration, mass flow rate, and dimensions of the flow cell. In case active analytic concentration is the only unknown variable, this can be measured without prior knowledge of the binding kinetics.[34]

5.2. *Limitation of Chemical Sensors and Suggestions*

In chemical sensor applications, the detection limit can be defined as the concentration of a target analyte generating a sensor signal three times greater than the standard deviation of noise from a blank. This simple definition, used to calculate the instrumental detection limit and is very useful for comparing the performance of different instrumental methods. This definition, however, poses two main problems. The first one is that it assumes chemical measurements have a normal and homogenous distribution over the measured concentration range, and second is that it only focuses on the probability for false-positives, i.e., the likelihood of obtaining a positive hit for a substance when truly it is not present. The second problem is particularly problematic because while the detection limit definition results in a very low probability for false-positives, its probability for true positives (the likelihood of obtaining a positive hit for a substance when it is truly present) is a low 0.50. In other words, on average, only 50% of all substances with a concentration equal to the detection limit will be detected. This is clearly unacceptable, especially if one believes that the detection limit should be the lowest concentration that can be reliably detected. The term "reliably detected" can be quantitatively described as having an acceptable true-positive probability (TPP) and false-positive probability (FPP).

Receiver operator characteristic (ROC) curve method was suggested to solve the above problem. This method can enable determining the detection limit of dichotomous sensors such as the ion mass spectroscopy (IMS) device. The limits of sensor detection tested by ROC lines are very reliable because they can be operated on numerical and repetitive sensor analysis under realistic operational circumstances. Also ROC lines are very good at measuring and explaining how this operating circumstances and processing or sampling environments can affect a sensor's operational detection ability.[35]

5.3. *Outlooks of Disposable Chemical Sensor/Biosensor*

In recent times, rapid progress has been made in cellulose-paper-based biosensors as well as chemical sensors. By hybridizing and covalent bonding of carbon nanotubes as well as functional metal oxides with cellulose substrates, the potential and feasibility of environment-friendly, cost-effective, and disposable sensor devices with good sensing performance was investigated. However, research and development of the hybrid materials in conjunction with sensor devices is necessary to move forward to real applications. Some successful paper-based biosensors and chemical sensors have been developed. However, further investigations should be done for long-term stability in harsh environments as well as industrial detection protocols to provide complementary sensing information. Table 5.4 shows lists the recently published details on paper-based electrochemical biosensors.[36]

Recently, a paper-based screen-printed biosensor was reported for the detection of ethanol in beer for quality control of products in food processing. Using a screen-printed electrode (SPE), paper-based electrochemical biosensor is able to successfully detect ethanol in different beer typologies as a disposable device. The research developed a nanocomposite formed by carbon black and Prussian blue nanoparticles as an electrocatalyst to detect the hydrogen peroxide generated by the enzymatic reaction between alcohol oxidase and ethanol.[37] After optimizing the analytical parameters, such as pH, enzyme, concentration, and working potential, the developed biosensor allowed a facile quantification of ethanol, as shown in Fig. 5.31. Using this concept, paper-based biosensor can offer an affordable and sustainable option for food quality control and for the development of different electrochemical sensors and biosensors as well.

Table 5.4. Characteristics of paper-based electrochemical biosensors.[36]

Reference	Paper	Hydrophobic patterning	Working electrode	Electrode modification materials	Biorecognition element	Analyte
Dungchai (2009)[24]	Whatman #1	Photolithography	SPCE array	Prussian Blue	GOx, Lactate Oxidase, Uricase	Glucose, Lactate, Uric Acid
Nie (2010)[23]	Whatman #1	Photolithography or wax printing	SPCE	Ferricyanide	GOx	Glucose and Pb (II)
Jagadeesan (2012)[25]	Whatman #1	—	SPCE	PANI, ferricyanide in solution	Antibodies	Troponin
Tan (2012)[26]	Whatman #1	—	Commercial SPCE	—	GOx	Silver ions
Wu (2012)[27]	Filter paper strips	—	SWCNT	—	Antibodies	Neomycin
Määttänen (2013)[28]	Multilayer-coated recyclable paper	PDMS ink	Gold SPE	PEDOT	GOx	Glucose
Noiphung (2013)[29]	Whatman #1 and VF separation paper	Wax dipping	Commercial Prussian Blue SPCE	Prussian Blue	GOx	Glucose
Santhiago (2013)[30]	Whatman #1	Wax printing	Graphite-pencil	4-aminophenylboronic acid	GOx	Glucose
Zhao (2013)[31]	Whatman #1	Wax printing	SPCE array	Ferricyanide	GOx, Lactate Oxidase, Uricase	Glucose, Lactate, Uric Acid

(*Continued*)

Table 5.4.   (Continued)

Reference	Paper	Hydrophobic patterning	Working electrode	Electrode modification materials	Biorecognition element	Analyte
Kong (2014)[32]	Whatman #1	—	Commercial SPCE	Graphene, PANI,AuNPs	GOx	Glucose
Labroo (2014)[33]	Regular paper	Wax printing	Inkjet printed graphene	—	GOx, Lactate Oxidase, XO, ChOx	Glucose, Lactate, Xanthine, Cholesterol
Lawrence (2014)[11]	Whatman #1	—	Commercial SPCE	Ferrocene monocarboxylic acid	GOx	Glucose
Li (2014)[34]	Whatman #1	Wax printing	SPCE	AuNPs, MNO$_2$ nanowires	Antibodies	PSA
Ruecha (2014)[35]	Whatman #1	Wax printing	SPCE	Graphene, polyvinylpyrrolidone and PANI	ChOx	Cholesterol
Wu (2014)[36]	Whatman #1	SU-8 photoresist	SPCE	Graphene oxide, chitosan and glutaraldehyde	Antibodies	Cancer biomarkers
Yang (2014)[37]	Whatman #1	—	Commercial SPCE	AgNPs	GOx	Glucose
Ge (2015)[38]	Chromatographic	Wax printing	SPCE	AuNPs, graphene, IL	Concanavalin A	K-562 cells
Ge (2015)[39]	Chromatographic	Wax printing	SPCE	AuNPs	Folic acid	K-562 cells
Li (2015)[40]	Whatman #1	Wax printing	SPCE	—	Labeled DNA probe	Viral DNA
Nantaphol (2015)[41]	Whatman #1	Wax printing	Boron-doped diamond	AgNPs	ChOx	Cholesterol

Wu (2015)[42]	Whatman #1	Wax printing	SPCE	3-aminopropyl-dimethod-xysiloxane, NAD^+ and ferricyanide	ADH	Ethanol
Fischer (2016)[43]	Whatman #1	Wax printing	SPCE	Chitosan	GOx	Glucose
Li (2016)[44]	Whatman #1	Not specified	Pencil drawn graphitic layers	Ferrocenecarboxylic acid	GOx	Glucose

ADH—alcohol dehydrogenase; AgNPs–silver nanoparticles; AuNPs–gold nanoparticles; ChOx–cholesterol oxidase; GOx–glucose oxidase; IL–ionic liquid; NAD^+–nicotinamide adenine dinucleotide; PANI–polyaniline; PEDOT–poly-3,4-ethylenedioxythiophene; PDMS–polydimethysiloxane; PSA–prostate protein antigen; SPCE–screen-printed carbon electrode; SPE–screen-printed electrode; ssDNA–single-stranded DNA; SWCNT–single-walled carbon nanotubes; XO–xanthine oxidase.

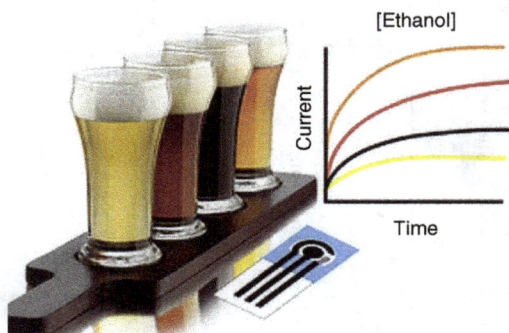

Fig. 5.31. Ethanol detection in commercial beers using paper sensor.[37]

References

1. J. M. Kim, S.-M. Chang, H. Muramatsu and K. Isao, The principles and applications of nano-diagnosis system for a nano-biosensor, *Korean J. Chem. Eng.* **28**(4), 987–1008 (2011).
2. N. J. Ronkainen, H. B. Halsall and W. R. Heineman, Electrochemical biosensors, *Chem. Soc. Rev.* **39**(5), 1747–1763 (2010).
3. K. R. Rogers, Principles of affinity-based biosensors, *Mol. Biotechnol.* **14**(2), 109–129 (2000).
4. M. Keusgen, Biosensors: New approaches in drug discovery, *Naturwissenschaften* **89**(10), 433–444 (2002).
5. T. Vo-Dinh and B. Cullum, Biosensors and biochips: advances in biological and medical diagnostics, *Fresenius' J. Anal. Chem.* **366**(6–7), 540–551 (2000).
6. T. Takeuchi, M. Yoshida, Y. Kabasawa, R. Matsukawa, E. Tamiya and I. Karube, Time-resolved fluorescence receptor assay for benzodiazepines, *Anal. Lett.* **26**(7), 1535–1545 (1993).
7. M. I. Song, K. Iwata, M. Yamada, K. Yokoyama, T. Takeuchi, E. Tamiya and I. Karube, Multisample analysis using an array of microreactors for an alternating-current field-enhanced latex immunoassay, *Anal. Chem.* **66**(6), 778–781 (1994).
8. K. Yokoyama, K. Ikebukuro, E. Tamiya, I. Karube, N. Ichiki, and Y. Arikawa, Highly sensitive quartz crystal immunosensors for multisample detection of herbicides, *Anal. Chim. Acta* **304**(2), 139–145 (1995).
9. K. Yagiuda, A. Hemmi, S. Ito, Y. Asano, Y. Fushinuki, C. Y. Chen and I. Karube, Development of a conductivity-based immunosensor for sensitive detection of methamphetamine (stimulant drug) in human urine, *Biosens. Bioelectron.* **11**(8), 703–707 (1996).
10. I. Karube and N. Keijiro, Immobilized cells used for detection and analysis, *Curr. Opin. Biotechnol.* **5**(1), 54–59 (1994).

11. H. Nakamura, Y. Hirata, Y. Mogi, S. Kobayashi, K. Suzuki, T. Hirayama and I. Karube, A simple and highly repeatable colorimetric toxicity assay method using 2,6-dichlorophenolindophenol as the redox color indicator and whole eukaryote cells, *Anal. Bioanal. Chem.* **389**(3), 835–840 (2007).

12. I. Karube, T. Matsunaga, S. Tsuru and S. Suzuki, Continuous hydrogen production by immobilized whole cells of Clostridium butyricum, *Biochim. Biophys. Acta (BBA)-General Subjects* **444**(2), 338–343 (1976).

13. T. Takeuchi, K. Yokoyama, K. Kobayashi, M. Suzuki, E. Tamiya, I. Karube and K. Utsunomiya, Photosynthetic activity sensor for microalgae based on an oxygen electrode integrated with optical fibres, *Anal. Chim. Acta* **276**(1), 65–68 (1993).

14. D. R. Thévenot, K. Toth, R. A. Durst and G. S. Wilson, Electrochemical biosensors: recommended definitions and classification, *Biosens. Bioelectron.* **16**(1), 121–131 (2001).

15. A. Koyun, E. Ahlatcioglu and Y. K. Ipek, Biosensors and their principles. In *A Roadmap of Biomedical Engineers and Milestones*, InTech: Rijeka, Croatia (2012).

16. N. Couniot, L. A. Francis and D. Flandre, A 16×16 CMOS capacitive biosensor array towards detection of single bacterial cell, *IEEE Trans. Biomed. Circuit Syst.* **10**(2), 364–374 (2015).

17. C. Berggren, P. Stalhandske, J. Brundell and G. Johansson, A feasibility study of a capacitive biosensor for direct detection of DNA hybridization, *Electroanalysis* **11**(3), 156–160 (1999).

18. A. Qureshi, Y. Gurbuz and J. H. Niazi, Capacitive aptamer–antibody based sandwich assay for the detection of VEGF cancer biomarker in serum, *Sensors Actuators B: Chem.* **209**, 645–651 (2015).

19. N. Tekaya, O. Saiapina, H. B. Ouada, F. Lagarde, P. Namour, H. B. Ouada and N. Jaffrezic-Renault, Bi-Enzymatic conductometric biosensor for detection of heavy metal ions and pesticides in water samples based on enzymatic inhibition in arthrospira platensis, *J. Environ. Prot.* **5**(5), 441–453 (2014).

20. S. Pal and E. C. Alocilja, Electrically active polyaniline coated magnetic (EAPM) nanoparticle as novel transducer in biosensor for detection of Bacillus anthracis spores in food samples, *Biosens. Bioelectron.* **24**(5), 1437–1444 (2009).

21. A. Tarasov, D. W. Gray, M.-Y. Tsai, N. Shields, A. Montrose, N. Creedon, P. Lovera, A. O'Riordan, M. H. Mooney and E. M. Vogel, A potentiometric biosensor for rapid on-site disease diagnostics, *Biosens. Bioelectron.* **79**, 669–678 (2016).

22. M. L. Johnson, J. Wan, S. Huanga, Z. Chenga, V. A. Petrenko, D.-J. Kim, I.-H. Chen, J. M. Barbaree, J. W. Hong and B. A. Chin, A wireless biosensor using microfabricated phage-interfaced magnetoelastic particles, *Sensors Actuators A: Phys.* **144**(1), 38-47 (2008).

23. G.-J. Chee, N. Yoko, I. Kazunori and K. Iaso, Biosensor for the evaluation of biochemical oxygen demand using photocatalytic pretreatment, *Sensors Actuators B: Chem.* **80**(1), 15–20 (2001).

24. A. Tereshchenko, M. Bechelany, R. Viter, V. Khranovskyy, V. Smyntyna, N. Starodub and R. Yakimova, Optical biosensors based on ZnO nanostructures: Advantages and perspectives: A review, *Sensors Actuators B: Chem.* **229**, 664–677 (2016).

25. M. Dahmardeh, S. Sheybanifar, M. Gharooni, M. Janmaleki and M. Abdolahad, Acoustic wave based biosensor to study electroacoustic based detection of progressive (SW-48) colon cancer cells from primary (HT-29) cells, *Sensors Actuators A: Phys.* **233**, 169–175 (2015).

26. A. Poturnayova, I. Karpisova, G. Castillo, G. Mező, L. Kocsis, A. Csámpai, Z. Keresztes and T. Hianik, Detection of plasmin based on specific peptide substrate using acoustic transducer, *Sensors Actuators B: Chem.* **223**, 591–598 (2016).

27. Y. Maria, S. Bhand and B. Danielsson, The enzyme thermistor — A realistic biosensor concept. A critical review, *Anal. Chim. Acta* **766**, 1–12 (2013).

28. J.-H. Kim, S. Mun, H.-U Ko, G. Yun and J. Kim, Disposable chemical sensors and biosensors made on cellulose paper, *Nanotechnology* **25**, 092001 (2014).

29. S. Ge, L. Ge, M. Yan, X. Song, J. Yu and J. Huang, A disposable paper-based electrochemical sensor with an addressable electrode array for cancer screening, *Chem. Commun.* **48**, 9397–9399 (2012).

30. L. Rivas, M. Medina, A. de la Escosura-Muñiz and A. Merkoçi, Improving sensitivity of gold nanoparticles-based lateral flow assays by using wax-printed pillars as delay barriers of microfluidics, *Lab Chip* **14**, 4406–4414 (2014).

31. C. Parolo, A. de la Escosura-Muñiz and A. Merkoçi, Enhanced lateral flow immunoassay using gold nanoparticles loaded with enzymes, *Biosens. Bioelectron.* **40**, 412–416 (2013).

32. E. Morales-Narváez, H. Golmohammadi, T. Naghdi, H. Yousefi, U. Kostiv, D. Horak, N. Pourreza and A. Merkoçi, Nanopaper as an optical sensing platform, *ACS Nano* **9**, 7296–7305 (2015).

33. B. Heli, E. Morales-Narváez, H. Golmohammadi, A. Ajji and A. Merkoçi, Modulation of population density and size of silver nanoparticles embedded in bacterial cellulose via ammonia exposure: Visual detection of volatile compounds in a piece of plasmonic nanopaper, *Nanoscale* **8**, 7984–7991 (2016).

34. P. M. Richalet-Sécordel, N. Rauffer-Bruyere, L. L. Christensen, B. Ofenloch-Haehnle, C. Seidel and M. H. Van Regenmortel, Concentration measurement of unpurified proteins using biosensor technology under conditions of partial mass transport limitation, *Anal. Biochem.* **249**(2), 165–173 (1997).

35. C. G. Fraga, A. M. Melville and B. W. Wright, ROC-curve approach for determining the detection limit of a field chemical sensor, *Analyst* **132**(3), 230–236 (2007).

36. C. M. Silveira, T. Monteiro and M. G. Almeida, Biosensing with paper-based miniaturized printed electrodes — A modern trend, *Biosensors* **6**, 51 (2016).

37. S. Cinti, M. Basso, D. Moscone and F. Arduini, A paper-based nanomodified electrochemical biosensor for ethanol detection in beers, *Anal. Chim. Acta* **960**, 123–130 (2017).

Index